# ALGEBRA EXAMPLES

## POWERS AND LOGARITHMS 3

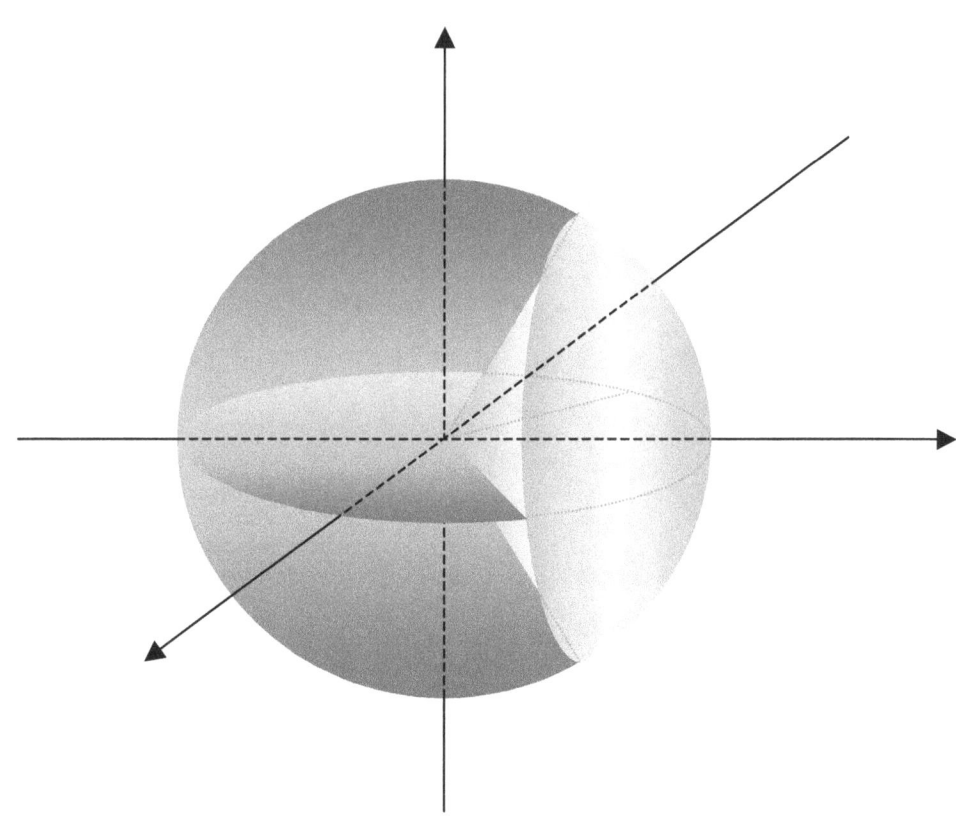

Seong R. KIM

# Dear students:

Students need the best teacher, so you need examples, because examples are the best teacher. All the examples here are fully worked, and explain **how** the basic and essential tools in math are made, together with **what** they are, **how** they work, and **how** to work with them. Such tools include numbers, formulas, identities, equations, laws, etc.

Examples here begin with easy ones, of course. Covering every meter and yard properly, we can cover thousands of miles and kilometers. And it is particularly the case in math.

Of those examples therefore, some might even look too easy for you. It's not that easy though, to come up with those examples. Anyways, the bigger and the taller the tree, the deeper and the stronger the root.

Doing math, we work with ideas and run ideas, because every thing in math is an idea. A number is an idea, for instance, and the same is true for a line or circle, too. And putting ideas together, we build another, which becomes the base or an element of another, and each is connected. And that's the way your math grows. So you get to build a circuit, and sometimes, need to fill the gap or repair the circuit so that you get the sense of it.

So your calculation runs properly, and you get the problem solved.

The examples have been made and arranged so that they get tougher (or sometimes easier for some reason) as you proceed with them. In particular, similar examples with some variations are strategically repeated so that you can get the ideas or the tools tricky or complicated, and can get them mastered.

This book is however, nothing but a bunch of examples until you get it powered.   How then, to get it powered, and make it run and work for you?

Just read it, and then, do each example in writing. And it is important to note that you do it in **your** writing. Just watching someone doing it, you just only feel that you can do it. If you do it, you can do it, but if you don't, we can hardly. It's a cliché, of course, but is always true that knowing is one thing and doing is another.

I've been helping students grow, take care of, and run their own math. The area covers algebra and geometry for high school or college students, and is especially for equations (for unknowns or curves), functions, and their graphs, which are the basic elements in calculus, which's been the core of my interest from my early age in high school.

Of my students, some are quite poor in math, and thus, are afraid of or hate math, some require special education because of exceptional intelligence, some are smart enough, some are naïve and diligent, some are clever but lazy, and most behave in general. All the students are badly after though, one thing in common: a strong and secure math skill. It is of course, the prime objective of my work, and I'm always happy to and eager to help them achieve it. The problem was however, that many of them wanted it to be purchased. And the question is, can we buy it?

We can buy the means, of course. And a solid math skill is feasible, too. We know however, we can't buy love, and the same is true for the math skill, too. It's not what we can buy or sell, and not what we can give or take. It is however, what we can grow, and need to grow. Your math grows as much as you grow and take care of it. So does mine.

What math then, do students most often do or use in high schools or colleges?

It is algebra and geometry. What algebra though?

Elementary algebra, of course
Doing the algebra, we work with numbers (many in kinds), constants, variables, ratios, rates, expressions, equations, inequalities, functions, identities, formulas, laws, etc., together with signs and symbols. And if we want to do algebra properly, we want to know their natures and how they mingle with each other.

So studying math ideas or tools, you want to know **what** they are, **how** they work, and **how** to work with them or **what** to do with them.     What then, about the geometry?

Basically, the geometry has much to do with shapes, positions, and angles. The shapes begin with triangles and circles, and move on to rectangles, squares, parallelograms or rhombuses, trapezoids, tetragons, other polygons, polyhedrons, etc.

Doing the geometry, too, though, we need to do the algebra stated above. So it is analytic geometry, often called coordinate geometry, too. And doing it, we can specify positions using coordinates. So in the geometry, basically, we work with graphs. Putting a math idea in a graph, we can not only effectively think about it but actually see it, too, and therefore, can efficiently work with it. What idea then, is it?

The idea begins with a point, line, parabola, circle, ellipse, and hyperbola, called a conic section or basic curve, and then, moves on to other curves, planes, surfaces, volumes, and other objects in various dimensional spaces, together with vectors.

And using an angle, we can specify an amount of turn or change in direction.

So learning, using, or applying those ideas or math tools, we get to solve problems.

And this book can help. It can help learn them, and use them so that you can navigate to find solutions to problems. And in particular, it can help come up with answers to those **what**s and **how**s stated above. So it can help you grow and run your own math, and thus, can help achieve your solid math skill.

It is however, not a magic book giving you a math skill of high caliber overnight. And it can have many mistakes, too. There is no magic, and math is full of facts and ideas. And it is after all, not me and not your teacher but you who put together some of those facts and ideas, and understand it. Putting facts and ideas together, understanding it, and taking care of what you have learned, you grow your math. And this book can help.

This is a book of examples designed to help you grow your math, and assumes that you are a real beginner. This book requires though, time and effort, the amount of which need to be substantial, too, but will be worth it. That's because you want a substantial achievement, and will get it. And probably, you will get to see this book helping you get there much faster than expected. And then, you will get to see the way math runs.

In math, everything is an idea. So is a problem. And solving it, we put it many different ways. For instance, while expanding or reducing it, or modifying or converting it, we keep searching for the solution, approaching the solution, and eventually, can get there. So don't look for the solution outside the problem. The solution is inside the problem if the problem is properly made.

If it is not, no solution is the solution. And in fact, it is often the case a problem itself is the solution. We can put a problem in many different ways, and eventually, can end up with the solution.   How come then, is the solution no other than the problem?

For instance, the solution to $3232 \div 101$ is 32.   And we can put it this way:

$$3232 \div 101 = \frac{3232}{101} = \frac{32 \times 101}{101} = \frac{32}{1} = 32 \ \Rightarrow 3232 \div 101 = 32.$$

And we can get this, too: $32 \Rightarrow 3232 \div 101$.   How?

$$32 = \frac{32}{1} = \frac{32 \times 101}{101} = \frac{3232}{101} = 3232/101 = 3232 \div 101. \quad \text{Too easy?}$$

For another instance, the solution to $ax^2 + bx + c = 0$ is: $x = \frac{-b \pm \sqrt{b^2 - 4ac}}{2a}$, which is called the quadratic formula.   How come then, is the solution no other than the problem?

We can put it this way:

$$x = \frac{-b \pm \sqrt{b^2 - 4ac}}{2a} \Rightarrow 2ax = -b \pm \sqrt{b^2 - 4ac} \Rightarrow 2ax + b = \pm \sqrt{b^2 - 4ac}$$

$$\Rightarrow (2ax + b)^2 = b^2 - 4ac \Rightarrow 4a^2x^2 + 4abx + b^2 = b^2 - 4ac$$

$$\Rightarrow 4a^2x^2 + 4abx = -4ac \Rightarrow ax^2 + bx = -c \Rightarrow ax^2 + bx + c = 0.$$

And we can get this, too: $ax^2 + bx + c = 0 \Rightarrow x = \frac{-b \pm \sqrt{b^2 - 4ac}}{2a}$.   How?

$$ax^2 + bx + c = a(x^2 + \tfrac{b}{a} x) + c = a(x^2 + \tfrac{b}{a}x + \tfrac{b^2}{4a^2} - \tfrac{b^2}{4a^2}) + c = a(x^2 + \tfrac{b}{a}x + \tfrac{b^2}{4a^2}) - \tfrac{b^2}{4a} + c$$

$$= a(x + \tfrac{b}{2a})^2 - \tfrac{b^2 - 4ac}{4a} = 0 \Rightarrow a(x + \tfrac{b}{2a})^2 = \tfrac{b^2 - 4ac}{4a} \Rightarrow (x + \tfrac{b}{2a})^2 = \tfrac{b^2 - 4ac}{4a^2} \Rightarrow x + \tfrac{b}{2a} = \pm\sqrt{\tfrac{b^2 - 4ac}{4a^2}}$$

$$\Rightarrow x = -\tfrac{b}{2a} \pm \tfrac{\sqrt{b^2 - 4ac}}{2a} = \tfrac{-b \pm \sqrt{b^2 - 4ac}}{2a} \Rightarrow x = \tfrac{-b \pm \sqrt{b^2 - 4ac}}{2a}.$$

And we call the set of processes above, algebra.

So if a problem is well defined, that is, if it makes sense, we should be able to get it solved the way below:

**A problem $\Rightarrow$ ... $\Rightarrow$ ... $\Rightarrow$ the solution**, and thus: **the problem $\Rightarrow$ the solution**.

So solving a problem, we put it many different ways so that we can get to the solution.

And that's the way, math runs.

May your math run very well.

Seong R. Kim

B.S. Math. Michigan Tech. Univ.    M.S. Math. Rensselaer Polytechnic Institute

# Notes:

This book is the continuation of the book, **ALGEBRA EXAMPLES POWERS AND LOGARITHMS 2**, which is about math ideas called logarithms.
So if you have not studied the book above, you may want to begin with the book.

Why logarithms though?

They are no other than exponents, and are in fact, for your algebra. Doing algebra, you need to know what logarithms are about as well as powers and exponents, how they work so that you can work with them. Why algebra though?

It's simply because you need to *solve problems*. Algebra connects problems to solutions. Algebra only can in fact, actually get you the ones you are always busy finding when taking exams, tests, and quizzes, and doing homework, too, of course. You've got to do algebra to get the very one you want so that you can put it down on *your answer sheet*. With algebra skill, together with your creativity, you can actually solve problems.

And this book is in fact, for your skill of algebra, and you will grow it through examples. Some examples may look too easy or too hard. It all depends on your skill of algebra. Whatever your skill may be though, you can grow yours if you follow the steps in each example. Each is detailed so that you can learn those tools fast, and increase your caliber quickly as well as properly.

And this book explains what logarithms are about and how to manipulate them, that is, how to change or alter, convert, or modify those so that you can come up with the ones that you need. The ones are solutions, of course. And that's what this book is about.

This book does not just explain though. But it helps follow steps to the solutions, too, and thus, helps you do calculations with logarithms as well as powers and exponents so that you can actually do the calculation work doing those manipulations above.

With strong algebra, you can learn things in math fast, and can do problems very well, too, of course. And this book is for your skill of log algebra, which is on expressions with logarithms. And there are two more books connected to this book, and the two are as follows:

**ALGEBRA EXAMPLES POWERS AND LOGARITHMS 2**

**ALGEBRA EXAMPLES POWERS AND LOGARITHMS 1**

And also, all the basics on powers and logarithms and all the ideas contained in this book and the two books above are covered in one book, too. And the book is as follows:

**ALGEBRA EXAMPLES POWERS AND LOGARITHMS**

So either way, the books will get you not only powers and logarithms but enhancement of your algebra, too. You will thus, soon be able to control powers and logarithms, that is, change or alter, convert, or modify those math expressions so that you can get to the solutions fast. And you will learn them all through examples detailed so that your math can run not only properly but fast enough, too.

# Contents

## In POWERS AND LOGARITHMS 3

# The Preview of the Contents

## In POWERS AND LOGARITHMS 1

# The Preview of the Contents

## In POWERS AND LOGRAITHMS 2

$$(x+y)^2 = x^2 + 2xy + y^2.$$

$$(x+y)^3 = x^3 + 3x^2y + 3xy^2 + y^3.$$

$$(x+y)(x-y) = x^2 - y^2.$$

$$(x+y)(x^2 - xy + y^2) = x^3 + y^3.$$

$$(x^2 + xy + y^2)(x^2 - xy + y^2) = x^4 + x^2y^2 + y^4.$$

$$(x+a)(x+b) = x^2 + (a+b)x + ab.$$

$$(ax+b)(cx+d) = acx^2 + (ad+bc)x + bd.$$

$$(x+a)(x+b)(x+c) = x^3 + (a+b+c)x^2 + (ac+bc+ca)x + abc.$$

$$(a+b+c)^2 = a^2 + b^2 + c^2 + 2(ab+bc+ca).$$

$$(a+b+c)(a^2 + b^2 + c^2 - ab - bc - ca) = a^3 + b^3 + c^3 - 3abc.$$

Suppose both $a$ and $b \neq 0$, and both $m$ and $n$ are integers. Then, we get:

0. $a^m a^n = a^{m+n}$        1. $a^m / a^n = \dfrac{a^m}{a^n} = a^{m-n}$        2. $(a^m)^n = a^{mn}$

3. $(ab)^n = a^n b^n$        4. $(a/b)^n = \left(\dfrac{a}{b}\right)^n = a^n / b^n = \dfrac{a^n}{b^n}$

Suppose both $a$ and $b > 0$, and $m$ and $n$ both are integers nonzero. Then, we get:

0.1. $a^{\frac{1}{n}} b^{\frac{1}{n}} = (ab)^{\frac{1}{n}}$.        1.1. $\dfrac{a^{\frac{1}{n}}}{b^{\frac{1}{n}}} = \left(\dfrac{a}{b}\right)^{\frac{1}{n}}$.        2.1. $(a^{\frac{1}{n}})^m = (a^m)^{\frac{1}{n}}$.

3.1. $(a^{\frac{1}{n}})^{\frac{1}{m}} = a^{\frac{1}{mn}} = (a^{\frac{1}{m}})^{\frac{1}{n}}$.        3.2. $(a^{mp})^{\frac{1}{np}} = (a^m)^{\frac{1}{n}}$, where $p$ is a nonzero integer.

1.   Suppose $M$, $N$, and $b > 0$, but $b \neq 1$, and we have: $A = \log_b M$, and $B = \log_b N$.
Then, we get: $A - B = \log_b M - \log_b N = \log_b \frac{M}{N}$.

2.   Suppose that $M$ and $b > 0$, but $b \neq 1$, and that we have: $E = \log_b M$.
Then, we get: $PE = P \log_b M = \log_b M^P$.

3.   Suppose that $a$, $b$, $C$, and $D > 0$, but $a$ and $b \neq 1$, and that we have: $\log_a C = \log_b D$.
Then, we get: $\log_a C = \log_b D = \log_{ab} CD$.

4.   Suppose that $a$, $b$, $C$, and $D > 0$, but $a$ and $b \neq 1$, and that we have: $\log_a C = \log_b D$.
Then, we get: $\log_a C = \log_b D = \log_{\frac{a}{b}} \frac{C}{D} = \log_{\frac{b}{a}} \frac{D}{C}$.

5.   $\log_b b = 1$, and $\log_b 1 = 0$.        6.   $\log_b A = \dfrac{\log_c A}{\log_c b}$.

7.   $\log_b A = \dfrac{1}{\log_A b}$.

**Note:**

The drawings or graphs in this book are not exact, and are approximate or conceptual ones.

| | |
|---|---|
| $\in$ | "$a \in B$" means that $a$ belongs to $B$.<br>"$p, q,$ **and** $r \in W$" means that $p, q,$ and $r$ belong to $W$. |
| $\Rightarrow$ | "$A \Rightarrow B$." means that $A$ implies $B$. |
| $\equiv$ | $A \equiv B$ means that $A$ and $B$ are identical to each other. |
| $\neq$ | $A \neq B$ means that $A$ is not equal to $B$. |
| $\lvert A \rvert$ | The magnitude of $A$. For instance, $\lvert -1 \rvert = \lvert 1 \rvert = 1$. |
| $\therefore$ | Therefore |
| $\Leftrightarrow$ | "$A \Leftrightarrow B$" means "If $A$ then $B$." and "If $B$ then $A$."<br>We can read $A \Leftrightarrow B$ as "$A$ if and only if $B$."<br>In such a case, we can say that $A = B$. |
| $\Delta x$ and $\Delta y$ | Suppose that $(x_1, y_1)$ and $(x_2, y_2)$ are two points in the $x$-$y$ plane. Then, we get either of the two below.<br><br>$\Delta x = x_2 - x_1$, and $\Delta y = y_2 - y_1$.<br><br>$\Delta x = x_1 - x_2$, and $\Delta y = y_1 - y_2$. |

**Distance Formula**

Suppose that $d$ is the distance between two points $(x_1, y_1)$ and $(x_2, y_2)$ in the $x$-$y$ plane. Then, we get $d^2 = (\Delta x)^2 + (\Delta y)^2$.

## Examples 1 on Logarithms

Assuming $a > 0$ and $\neq 1$, simplify the following expressions:

0. $\log_a a$

1. $\log_a 1$

2. $\log_a a^k$

3. $\log_3 27$

4. $\log_{\sqrt{3}} 27$

5. $\log_{\sqrt{3}} \frac{1}{27}$

Do log arithmetic below.

0.6. $\log_3 \frac{1}{27} + \log_3 27$

0.7. $\log_{15} \frac{1}{27} + \log_{15} 27$

0.8. $\log_2 96 - \log_2 3$

0.9. $\log_3 54 - \log_3 2$

Assuming $p$, $Q$, and $c > 0$, but $\neq 1$, find $x$ in each of the equations below.

0.A. $\log_p Q = \dfrac{\log_x Q}{\log_c p}$

0.B. $\log_p Q = \dfrac{1}{\log_x p}$

## Suggestions or Solutions
## To the Problems in the Examples 1

This set of examples provides some practice material on basic algebra on logs, so it should help get more familiar with manipulation of logs. And the same it true for all the other sets, too. That's simply because algebra matters.

Math begins with definitions, so let's begin this practice with giving a brief visit to the definition for logs, which is presented below.

- Suppose $A$ and $b$ are real and $> 0$, but $b \neq 1$. Then, $A = b^x \Leftrightarrow x = \log_b A$ where $x$ is real.

**0.** Let $x = \log_a a$. Then, we get: $a^x = a$ by the definition for logs.

Thus, $x = 1$. So we get: $\log_a a = 1$.

**1.** Let $x = \log_a 1$. Then, we get: $a^x = 1$ by the definition.

Thus, $x = 0$. So we get: $\log_a 1 = 0$.

**2.** Let $x = \log_a a^k$. Then, we get: $a^x = a^k$ by the definition.

Thus, $x = k$. So we get: $\log_a a^k = k$.

And in fact, we often get to use: $\log_a a = 1$, $\log_a 1 = 0$, and $\log_a a^k = k$ doing log algebra.

3.   Let $x = \log_3 27$.  Then, by the definition, we get: $3^x = 27 = 3^3$.  Thus, $x = 3$.

4.   We can get: $27 = 3^3 = \{(\sqrt{3})^2\}^3 = (\sqrt{3})^6$.

So we get: $\log_{\sqrt{3}} 27 = \log_{\sqrt{3}}(\sqrt{3})^6 = 6\log_{\sqrt{3}}\sqrt{3} = 6$.

And we can get it the way below, too:

$$\log_{\sqrt{3}} 27 = \frac{\log 27}{\log\sqrt{3}} = \frac{\log 3^3}{\log 3^{\frac{1}{2}}} = \frac{3\log 3}{\frac{1}{2}\log 3} = \frac{3}{\frac{1}{2}} = 6.$$

5.   $\log_{\sqrt{3}} \frac{1}{27} = \dfrac{\log\frac{1}{27}}{\log\sqrt{3}} = \dfrac{\log 3^{-3}}{\log 3^{\frac{1}{2}}} = \dfrac{-3\log 3}{\frac{1}{2}\log 3} = \dfrac{-3}{\frac{1}{2}} = \text{-}6.$

6.   $\log_3 \frac{1}{27} + \log_3 27 = \log_3(\frac{1}{27}\cdot 27) = \log_3 1 = 0.$

7.   $\log_{15} \frac{1}{27} + \log_{15} 27 = \log_{15}(\frac{1}{27}\cdot 27) = \log_{15} 1 = 0.$

8.   $\log_2 96 - \log_2 3 = \log_2 \frac{96}{3} = \log_2 32 = \log_2 2^5 = 5.$

9.   $\log_3 54 - \log_3 2 = \log_2 \frac{54}{2} = \log_3 27 = \log_3 3^3 = 3.$

**A.** Assuming $p$, $Q$, and $c > 0$, but are unequal to 1, find $x$ in $\log_p Q = \dfrac{\log_x Q}{\log_c p}$.

We have: $\log_b A = \dfrac{\log_c A}{\log_c b}$, so we get: $\log_p Q = \dfrac{\log_x Q}{\log_c p} \Rightarrow x = c.$

*If not quite sure of the idea behind the processes above, follow the steps below:*

We have a log identity where $\log_b A = \dfrac{\log_c A}{\log_c b}$.

In the identity above, $c$ can be any positive number other than 1.

So for instance, we can have: $\log_2 3 = \dfrac{\log_5 3}{\log_5 2} = \dfrac{\log_{21.2} 3}{\log_{21.2} 2} = \ldots$

Therefore, $\log_p Q = \dfrac{\log_x Q}{\log_c p} \Rightarrow x = c.$

We often need to put a log in a fractional form the way above.

So we may want to take $\log_p Q = \dfrac{\log_x Q}{\log_c p}$ as a log identity, and get used to it.

**B.** Assuming $p$ and $Q > 0$, but are not equal to 1, find $x$ in $\log_p Q = \dfrac{1}{\log_x p}$.

We have: $\log_p Q = \dfrac{\log_c Q}{\log_c p}$, so we get: $\dfrac{1}{\log_p Q} = \dfrac{\log_c p}{\log_c Q} = \log_Q p.$

Thus, we get: $\log_p Q = \dfrac{1}{\log_Q p}$. Therefore, $\log_p Q = \dfrac{1}{\log_x p} \Rightarrow x = Q.$

*If not quite sure of the idea behind the processes above, follow the steps below:*

Doing log algebra, we often need to make the base and antilog exchange their positions. How?

We can use a log identity where $\log_b A = \dfrac{1}{\log_A b}$, where both $A$ and $b > 0$, but $\neq 1$.

So we can immediately see that $\log_p Q = \dfrac{1}{\log_x p} \Rightarrow x = Q.$

Thus, the identity above can be a handy tool. So using it, we can readily make the base and antilog swap their positions.

Let's see though, how the identity above can hold. This proof is a bit different from the one covered earlier.

To begin with, we have: $\log_b A = \dfrac{\log_c A}{\log_c b}$, where both $A$ and $b > 0$, but $\neq 1$.

Taking the reciprocal of each side in the identity above, we get: $\dfrac{1}{\log_b A} = \dfrac{\log_c b}{\log_c A}$, which

equals $\log_A b$. So we get: $\dfrac{1}{\log_b A} = \log_A b.$

Therefore, $\log_b A = \dfrac{1}{\log_A b}$, which is often used, and can be a handy tool in log algebra.

**In short:**

We have: $\log_p Q = \dfrac{\log_c Q}{\log_c p}$, so we get: $\dfrac{1}{\log_p Q} = \dfrac{\log_c p}{\log_c Q} = \log_Q p.$

Thus, we get: $\log_p Q = \dfrac{1}{\log_Q p}$. Therefore, $\log_p Q = \dfrac{1}{\log_x p} \Rightarrow x = Q.$

## Examples 2 on Logarithms

In this set of examples, too, we are going to do some practice on basic algebra on logs so that we get more familiar with manipulation of logs. That's because algebra matters.

And math begins with definitions, so let's begin with a brief visit to the definition for logs, which is below.

- Suppose $A$ and $b$ are real and $> 0$, but $b \neq 1$. Then, $A = b^x \Leftrightarrow x = \log_b A$ where $x$ is real.

0.   Suppose now, $A$, $b$, $c$, and $d > 0$, but $b$ and $c \neq 1$.

Suppose also, $b = c^m$ where $m \neq 0$, and $A = d^n$.

Then, find $k$ for which $\log_b A = k \log_c d$.

1.   Assuming that $A$ and $b > 0$, but $b \neq 1$, find $k$ for which:

1.0.   $\log_{\frac{1}{b}} \frac{1}{A} = k \log_b A$

1.1.   $\log_{\frac{1}{b}} A = k \log_b A$

2.   Assuming that $a$, $b$, $c$, and $d$ all $> 0$, but $\neq 1$, simplify the following expressions:

2.0. $(\log_a b)(\log_b a)$

2.1. $(\log_c a)(\log_b c)(\log_a b)$

2.2. $(\log_a b)(\log_b c)(\log_c d)$

## Suggestions or Solutions
## To the Problem in the Example 0

Suppose $A$, $b$, $c$, and $d > 0$, but $b$ and $c \neq 1$.
Suppose also, $b = c^m$ where $m \neq 0$, and $A = d^n$.

Then, find $k$ for which $\log_b A = k \log_c d$.

We have: $A = d^n$, and $b = c^m$, where $m \neq 0$.

So $\log_b A = \frac{\log_c A}{\log_c b} = \frac{\log_c d^n}{\log_c c^m} = \frac{n\log_c d}{m\log_c c} = \frac{n\log_c d}{m} = \frac{n}{m}\log_c d = k \log_c d$.  Thus, $k = \frac{n}{m}$.

*If not quite sure of the idea behind the processes above, follow the steps below:*

This problem is nothing but a log equation for $k$.

Thus, we need to solve for $k$ the equation where $\log_b A = k \log_c d$.

Suppose now, $E$ is the equation $\log_b A = k \log_c d$.

Then, the equation $E$ has a condition that $b = c^m$ where $m \neq 0$, and $A = d^n$, so $E$ looks quite complicated. Such a condition however, should be taken for not a burden but a means. So we may want to take rather advantage of than offense at that.

Closely looking at the equation $E$, together with the condition, we can see that $b$ and $A$ can be removed from $E$. In other words, we can get a new equation that doesn't have $b$ and $A$, but is equivalent to the equation $E$.    How?

In the condition given, we have: $b = c^m$, and $A = d^n$.

So we can substitute $b$ with $c^m$, and $A$ with $d^n$ in the equation $E$.

Prior to substitutions however, we may want to modify the left hand side of $E$ so that it is in a fractional form where both the numerator and denominator have the same base $c$.

We can do such a modification later, though. It doesn't matter if we do the substitutions first, and then, put the left hand side in such a fractional form.

Why then, such a fractional form?

That's because the right hand side is a log to base $c$.
In math, consistency matters.

Let's now, begin with the left hand side of $E$ where $\log_b A = k \log_c d$.

Using a log identity, $\log_t s = \frac{\log_u s}{\log_u t}$, we can set: $\log_b A = \frac{\log_c A}{\log_c b}$, so we get: $\frac{\log_c A}{\log_c b} = k \log_c d$.

Let's next, move on to the substitutions.
Besides, after each substitution, we can use another log identity, where $\log_t s^w = w \log_t s$.

Then, we get:

$A = d^n \Rightarrow \log_c A = \log_c d^n = n \log_c d \Rightarrow \log_c A = n \log_c d$.

$b = c^m \Rightarrow \log_c b = \log_c c^m = m \log_c c = m$, since $\log_c c = 1$. Thus, $\log_c b = m$.

Now, putting threads together, we get: $\frac{\log_c A}{\log_c b} = \frac{n \log_c d}{m} = \frac{n}{m} \log_c d$, since $m \neq 0$.

Thus, we get: $\log_b A = \frac{\log_c A}{\log_c b} = \frac{n}{m} \log_c d = k \log_c d \Rightarrow k = \frac{n}{m}$.

So we can see that $\log_{c^m} d^n = \frac{n}{m} \log_c d$, which is often quite handy in log algebra.

For instance:

$\log_{27} 8 = \frac{3}{3} \log_3 2 = \log_3 2$, because $27 = 3^3$ and $8 = 2^3$.

$\log_{25} 32 = \frac{5}{2} \log_5 2$, because $32 = 2^5$ and $25 = 5^2$.

**In short:**

We have: $A = d^n$, and $b = c^m$, where $m \neq 0$.

So $\log_b A = \frac{\log_c A}{\log_c b} = \frac{\log_c d^n}{\log_c c^m} = \frac{n \log_c d}{m \log_c c} = \frac{n \log_c d}{m} = \frac{n}{m} \log_c d = k \log_c d$.  Thus, $k = \frac{n}{m}$.

## Suggestions or Solutions
## To the Problems in the Example 1

**1.0.    Assuming $A$ and $b > 0$, but $b \neq 1$, find $k$ for which $\log_{\frac{1}{b}} \frac{1}{A} = k \log_b A$.**

In the problem **1.0** above, we found that:

If $A = d^n$ and $b = c^m$ where $m \neq 0$, we get $\log_b A = \frac{n}{m} \log_c d$.

And we know: $\frac{1}{b} = b^{-1}$ and $\frac{1}{A} = A^{-1}$.

So we get:  $\log_{\frac{1}{b}} \frac{1}{A} = \frac{-1}{-1} \log_b A = \log_b A$.  Therefore, $k = 1$.

So we can see that $\log_{\frac{1}{b}} \frac{1}{A} = \log_b A$, which can be taken for another log identity.

**1.1.    Assuming $A$ and $b > 0$, but $b \neq 1$, find $k$ for which $\log_{\frac{1}{b}} A = k \log_b A$.**

We know: $\frac{1}{b} = b^{-1}$ and $A = A^1$.

So we get:  $\log_{\frac{1}{b}} A = \frac{1}{-1} \log_b A = -\log_b A$.  Therefore, $k = -1$.

We know: $-\log_b A = \log_b A^{-1} = \log_b \frac{1}{A}$.  Thus, we get: $\log_{\frac{1}{b}} A = \log_b \frac{1}{A} = -\log_b A$.

## Suggestions or Solutions
## To the Problems in the Example 2

**2.0.   Assuming that $a$ and $b$ both $> 0$, but $\neq 1$, simplify $(\log_a b)(\log_b a)$.**

The expression $(\log_a b)(\log_b a)$ can be taken for a product of reciprocals.

That is, $\log_a b = \frac{1}{\log_b a}$, and $\log_a b$ and $\log_b a$ can be taken for reciprocals to each other.

So we get: $(\log_a b)(\log_b a) = 1$.   How come then, logs can be reciprocals to each other?

Saying reciprocals, we mean a pair of numbers or expressions, the product of which is 1.

We know logs are exponents, we can use a number as an exponent, and if the number is not 0, it can have its reciprocal. So for instance, $2 = \log_3 9$ and $\frac{1}{2} = \log_9 3$.

Thus, $\log_3 9$ and $\log_9 3$ are reciprocals to each other.

We can put a log in a fractional form by means of a log identity as follows:

$$\log_p Q = \frac{\log_r Q}{\log_r p},$$ where $p$, $Q$, and $r > 0$, but both $p$ and $r \neq 1$.

So assuming that $c > 0$ but $\neq 1$, and setting $p = a$, $Q = b$, and $r = c$ in the identity above,

we get: $(\log_a b)(\log_b a) = \left( \dfrac{\log_c b}{\log_c a} \right)\left( \dfrac{\log_c a}{\log_c b} \right) = 1.$

Let's this time, set $p = a$, $Q = b$, and $r = b$, and see what happens.

Then, we get: $\log_a b = \dfrac{\log_b b}{\log_b a}$, and $\log_b a = \dfrac{\log_b a}{\log_b b}.$

So we get: $(\log_a b)(\log_b a) = \dfrac{\log_b b}{\log_b a} \cdot \dfrac{\log_b a}{\log_b b} = 1.$

In fact, we have: $\log_a b = \dfrac{\log_b b}{\log_b a} = \dfrac{1}{\log_b a}$, so we get: $(\log_a b)(\log_b a) = 1.$

**In short:**

$(\log_a b)(\log_b a) = \left(\dfrac{\log_c b}{\log_c a}\right)\left(\dfrac{\log_c a}{\log_c b}\right) = 1,$ or $(\log_a b)(\log_b a) = (\log_a b)\dfrac{\log_a a}{\log_a b} = 1.$

**2.1.  Assuming that $a$, $b$, and $c$ all $> 0$, but $\neq 1$, simplify $(\log_c a)(\log_b c)(\log_a b)$.**

Is the expression above a product of log reciprocals, too?

Eventually, it will be.

During the simplification process, we can see pairs of reciprocals.

So the product is going to be 1 again.

In fact, we frequently see such an expression in ordinary numbers, too.

For instance, $\left(\dfrac{2}{3}\right)\left(\dfrac{3}{5}\right)\left(\dfrac{5}{2}\right) = 2\left(\dfrac{1}{3}\right)3\left(\dfrac{1}{5}\right)5\left(\dfrac{1}{2}\right) = 2\left(\dfrac{1}{2}\right)3\left(\dfrac{1}{3}\right)5\left(\dfrac{1}{5}\right) = 1 \cdot 1 \cdot 1 = 1,$ and also,

$\left(\dfrac{2}{3}\right)\left(\dfrac{3}{5}\right)\left(\dfrac{5}{2}\right) = 2\left(\dfrac{1}{3}\right)3\left(\dfrac{1}{5}\right)\left(\dfrac{5}{2}\right) = 2\left(\dfrac{1}{5}\right)\left(\dfrac{5}{2}\right) = 2\left(\dfrac{1}{5}\right)5\left(\dfrac{1}{2}\right) = 2\left(\dfrac{1}{2}\right) = 1.$

Let's now, put in fractional forms the logs in the expression above, and simplify the result.    Assuming that $d > 0$, but $\neq 1$, we get:

$$(\log_c a)(\log_b c)(\log_a b) = \left(\frac{\log_d a}{\log_d c}\right)\left(\frac{\log_d c}{\log_d b}\right)\left(\frac{\log_d b}{\log_d a}\right) = \left(\frac{\log_d a}{\log_d b}\right)\left(\frac{\log_d b}{\log_d a}\right) = 1.$$

Putting a log in such a fractional form as above, we can use as the base $d$ any number that can be a base in a log. So it can be any positive number other than 1, and thus, we can produce the solution to this problem many ways. For instance:

$$(\log_c a)(\log_b c)(\log_a b) = \left(\frac{\log_2 a}{\log_2 c}\right)\left(\frac{\log_2 c}{\log_2 b}\right)\left(\frac{\log_2 b}{\log_2 a}\right) = 1.$$

$$(\log_c a)(\log_b c)(\log_a b) = \left(\frac{\log_3 a}{\log_3 c}\right)\left(\frac{\log_3 c}{\log_3 b}\right)\left(\frac{\log_3 b}{\log_3 a}\right) = 1.$$

**In short:**

$$(\log_c a)(\log_b c)(\log_a b) = \left(\frac{\log_4 a}{\log_4 c}\right)\left(\frac{\log_4 c}{\log_4 b}\right)\left(\frac{\log_4 b}{\log_4 a}\right) = 1.$$

**2.2.    Assuming that $a$, $b$, $c$, and $d$ all $> 0$, but $\neq 1$, simplify $(\log_a b)(\log_b c)(\log_c d)$.**

The expression above is not a product of log reciprocals, but has much to do with such a product.

During the simplification processes, we will see pairs of reciprocals, but the whole expression is not a product of pairs of reciprocals only.

We often see such an expression as above in ordinary numbers, too.

For instance, $\left(\dfrac{2}{3}\right)\left(\dfrac{3}{5}\right)\left(\dfrac{5}{4}\right) = 2 \cdot \dfrac{1}{3} \cdot 3 \cdot \dfrac{1}{5}\left(\dfrac{5}{4}\right) = \left(\dfrac{2}{5}\right)\left(\dfrac{5}{4}\right) = 2\left(\dfrac{1}{5}\right)5\left(\dfrac{1}{4}\right) = \dfrac{2}{4} = \dfrac{1}{2}$, and also,

$\left(\dfrac{2}{3}\right)\left(\dfrac{3}{5}\right)\left(\dfrac{5}{4}\right) = 2\left(\dfrac{1}{3}\right)3\left(\dfrac{1}{5}\right)5\left(\dfrac{1}{4}\right) = \dfrac{2}{4} = \dfrac{1}{2}$. Normally, we do it the way below, of course:

$\left(\dfrac{2}{\cancel{3}}\right)\left(\dfrac{\cancel{3}}{\cancel{5}}\right)\left(\dfrac{\cancel{5}}{4}\right) = \dfrac{2}{4} = \dfrac{1}{2}$.

Let's now, put in fractional forms the logs in the expression given above, and then, simplify the result.

We can use as the common base any number positive and unequal to 1 putting a log in a fractional form. So this time, let's use 3 as the base in each log in the fractional form.

Then, we get:

$(\log_a b)(\log_b c)(\log_c d)$

$= \left(\dfrac{\log_3 b}{\log_3 a}\right)\left(\dfrac{\log_3 c}{\log_3 b}\right)\left(\dfrac{\log_3 d}{\log_3 c}\right) = \left(\dfrac{\log_3 c}{\log_3 a}\right)\left(\dfrac{\log_3 d}{\log_3 c}\right) = \dfrac{\log_3 d}{\log_3 a} = \log_a d$.

We can do it the way below, too, of course:

$(\log_a b)(\log_b c)(\log_c d) = \left(\dfrac{\cancel{\log_3 b}}{\log_3 a}\right)\left(\dfrac{\cancel{\log_3 c}}{\cancel{\log_3 b}}\right)\left(\dfrac{\log_3 d}{\cancel{\log_3 c}}\right) = \dfrac{\log_3 d}{\log_3 a} = \log_a d$.

## Examples 3 on Logarithms

In this set of examples, too, we are going to do some practice on basic algebra on logs so that we get more familiar with manipulation of logs. That's because algebra matters.

0.     Let $a$, $b$, and $c > 0$, but $b \neq 1$. Then:

0.0.     Assuming that $a = b^k$ where $k = \log_b c$, show that $a = c$.

0.1.     Assuming that $\log_b a = a^k$ where $k = \frac{\log_b N}{\log_b a}$ where $N = \log_b c$, show that $a = c$.

1.     Let $a$, $b$, and $c > 0$, but $c \neq 1$.
Then, assuming $B = \log_c b$ and $A = \log_c a$, show that $a^B = b^A$.

2.     Let $A$ and $b > 0$, but $b \neq 1$.
Suppose also, $\log_b A = s + t$, where $s$ is an integer called the index, and $0 \leq t < 1$, where $t$ is called the mantissa. Then:

2.0.     Assuming that $b = 10$ in $\log_b A$ above, find the highest place value of $A$, and put the value in terms of the index $s$.

2.1.     Assuming that $X$ and $Y > 0$, and that $\log X$ and $\log Y$ have the same mantissa, but have different indices, show that both $X$ and $Y$ have the same sequence of digits, but the positions of the decimal points are different.

2.2.     Assuming that $\log X - \log Y = k$ where $k$ is an integer, find the ratio, $\frac{X}{Y}$.

## Suggestions or Solutions
## To the Problem 0 in the Example 0

**Let $a$, $b$, and $c > 0$, but $b \neq 1$. Then, assuming $a = b^k$ where $k = \log_b c$, show that $a = c$.**

We have $a = b^k \Leftrightarrow k = \log_b a$ by the definition for logs. Therefore, $a = c$.

*If not quite sure of the idea behind the processes above, follow the steps below:*

This example is nothing but a matter of the definition for logs.
We have the definition for logs as follows:

- Assuming $A$ and $b$ are real and $> 0$, but $b \neq 1$, we get: $A = b^x \Leftrightarrow x = \log_b A$.

Thus, by the definition, we get: $a = b^k \Leftrightarrow k = \log_b a$.

And in this problem, we have $a = b^k$ and $k = \log_b c$.   So we can see that $a = c$.

And also, we can come to the same conclusion that $a = c$ in another way.

Taking $\log_b$ of each side in the equality, $a = b^k$, we get: $\log_b a = \log_b b^k$.

We have this, too: $\log_b b^k = k \log_b b = k$ since $\log_b b = 1$.   So we get: $\log_b a = k$.

Besides, we are given: $k = \log_b c$.   So we get: $\log_b a = \log_b c$, and we can see that $a = c$.

In fact, the statement, "$a = b^k$ where $k = \log_b c \Rightarrow a = c$." is saying this: $c = b^{\log_b c}$.

We often use it doing log algebra.   For instance:
$$15 = 3^{\log_3 15}, \text{and } 3.9 = 1.2^{\log_{1.2} 3.9}, \text{ so } 15 = 3^{\log_3 15} = 4^{\log_4 15} = 7^{\log_7 15} = \ldots$$

Thus in general, assuming that $A$, $b$, $c$, and $d > 0$, but $b$, $c$, and $d$ are $\neq 1$, we get:

$$A = b^{\log_b A} = c^{\log_c A} = d^{\log_d A} = \ldots$$

## Suggestions or Solutions
## To the Problem 1 in the Example 0

**Let $a$, $b$, and $c > 0$, but $b \neq 1$.**

**Then, assuming $\log_b a = a^k$ where $k = \frac{\log_b N}{\log_b a}$ where $N = \log_b c$, show that $a = c$.**

To begin with, $k = \frac{\log_b N}{\log_b a} = \log_a N \Rightarrow k = \log_a N \Rightarrow N = a^k$ by the definition for logs.

Next, $\log_b a = a^k \Rightarrow \log_b a = N$ since $N = a^k$.

Also, we have: $N = \log_b c$.    Thus, we get: $\log_b a = \log_b c$.    Therefore, $a = c$.

*If not quite sure of the idea behind the processes above, follow the steps below:*

The expression looks quite complicated. Taking one step at a time though, we will see it can get actually quite simple.

Let's take care of $k = \frac{\log_b N}{\log_b a}$, first.

We have: $\frac{\log_b N}{\log_b a} = \log_a N$.   So we get: $k = \log_a N$.

Thus, we get: $N = a^k$ by the definition for logs.

Now, we have: $\log_b a = a^k$, and $N = a^k$.    So we get: $\log_b a = N$.

Besides, we have: $N = \log_b c$, also.    So we get: $\log_b a = \log_b c$.

Therefore, we can see that $a = c$.

## Suggestions or Solutions
## To the Problem in the Example 1

**Let $a$, $b$, and $c > 0$, but $c \neq 1$.**
**Then, assuming $B = \log_c b$, and $A = \log_c a$, show that $a^B = b^A$.**

We have: $B = \log_c b$, and $A = \log_c a$.

And setting $k = a^B$, we get:

$\log_c k = \log_c a^B = B \log_c a = (\log_c b)(\log_c a) = (\log_c b)A = A \log_c b$.

Thus, $\log_c k = A \log_c b = \log_c b^A$.

So we get: $k = b^A$.   Besides, we have already set: $k = a^B$.   Therefore, $a^B = b^A$.

*If not quite sure of the idea behind the processes above, follow the steps below:*

We have an assumption that $B = \log_c b$ and $A = \log_c a$.

Examining the assumption, we can see that $b$ in $\log_c b$ is used in $b^A$, and that $a$ in $\log_c a$ is used in $a^B$. Besides, $c$ is used as the base in each log in the assumption.

So taking $\log_c$ of each side of the equality, $a^B = b^A$, we can get the logs in the assumption.

Then, we can see better what we can do in the next step.

So let's begin with taking $\log_c$ of $a^B$, which is in the left hand side of the equality we have to show.

Setting $k = a^B$ for simplicity, and taking the log of it, we get: $\log_c k = \log_c a^B$.

And we know: $B \log_c a = \log_c a^B$.   So we get: $\log_c k = B \log_c a$.

Besides, we have: $B = \log_c b$, too.    So we get: $\log_c k = (\log_c b)(\log_c a)$.

Also, we have: $A = \log_c a$.    Thus, we get:

$\log_c k = (\log_c b)(\log_c a) = (\log_c b)A = A(\log_c b) = A \log_c b$.

Therefore, we get: $\log_c k = A \log_c b$.

Now, we know: $A \log_c b = \log_c b^A$.    So we get: $\log_c k = \log_c b^A$.

Thus, we get: $k = b^A$. And yet, we have already set: $k = a^B$ above.

Therefore, we get: $a^B = b^A$.

In sum, we have: $a^{\log_c b} = b^{\log_c a}$.  So $a$ and $b$ can exchange their positions.

For instance, $3^{\log_2 7} = 7^{\log_2 3}$.  Besides, we can have this, too:  $3^{\log_5 7} = 7^{\log_5 3}$.

**In short:**

We have: $B = \log_c b$, and $A = \log_c a$.

And setting $k = a^B$, we get:

$\log_c k = \log_c a^B = B \log_c a = (\log_c b)(\log_c a) = (\log_c b)A = A \log_c b$.

Thus, $\log_c k = A \log_c b = \log_c b^A$.

So we get: $k = b^A$.

Besides, we have already set: $k = a^B$.

Therefore, $a^B = b^A$.

## Suggestions or Solutions
## To the Problem 0 in the Example 5

Suppose that $A$ and $b > 0$, but $b \neq 1$, and that $\log_b A = s + t$, where $s$ is an integer called the index, and $0 \leq t < 1$, where $t$ is called the mantissa.

Then, assuming that $b = 10$ in $\log_b A$ above, find the highest place value in $A$, and put the value in terms of the index $s$.

By the definition for logs, $\log_b A = s + t \Leftrightarrow A = b^{s+t}$, where $s$ is an integer, and $0 \leq t < 1$.

Since $b = 10$, we can set: $\log A = s + t \Leftrightarrow A = 10^{s+t}$.

Thus, we have: $A = 10^{s+t} = 10^t 10^s$. So we get:

$0 \leq t < 1 \Rightarrow 10^0 \leq 10^t < 10^1 \Rightarrow 1 \leq 10^t < 10 \Rightarrow 10^s \leq 10^t 10^s < 10^{s+1} \Rightarrow 10^s \leq A < 10^{s+1}$.

Therefore, the highest place value in $A$ is $10^s$.

*If not quite sure of the idea behind the processes above, follow the steps below:*

When a base $b = 10$ in a log, we usually omit the base.

In that case, such a *log* is called a *common log*, the index is often called a characteristic, and also, we usually use a letter $c$ for indicating the characteristic, and a letter $m$ as the mantissa. Taking a common log of a number $A$ positive, we normally set:

$\log A = c + m$, where $c$ is an integer, often called a characteristic, and $0 \leq m < 1$.

Then, $10^c$ is the highest place value in the number $A$.

Now, in this example, we are going to see how the highest place value in a number is $10^c$. And math begins with definitions.

So doing problems in math, too, we may want to begin with definitions involved.

Thus, doing this problem, also, we may want to begin with $A = b^x \Leftrightarrow x = \log_b A$, which is the definition for logs. Of course, $x$ is real, and $A$ and $b$ both are $> 0$, but $b \neq 1$.

Besides, in this problem, even if $b = 10$, we will use $s$ instead of $c$.

That's simply because of consistency, and $s$ is used in the problem description.

And for the same reason, we will use $t$ as the mantissa.

So by the definition for logs, we can set:

$\log_b A = s + t \Leftrightarrow A = b^{s+t}$, where $s$ is an integer and $0 \leq t < 1$.

Now, in this problem, we have: $b = 10$, so we get: $\log A = s + t \Leftrightarrow A = 10^{s+t}$.

And we have: $0 \leq t < 1$, too.

Thus, we get: $0 \leq t < 1 \Rightarrow 10^0 \leq 10^t < 10^1 \Rightarrow 1 \leq 10^t < 10$.

Next, we have: $A = 10^{s+t}$, also, so we get: $A = 10^t 10^s$.

Thus, we get: $1 \leq 10^t < 10 \Rightarrow 10^s \leq 10^t 10^s < 10^{s+1} \Rightarrow 10^s \leq A < 10^{s+1}$.

Therefore, we can see that $10^s$ is the highest place value in the number $A$.    How come?

Suppose $w = 10^t$. Then, since $1 \leq 10^t < 10$, we get: $1 \leq w < 10$, so $w$ is 3.89 for instance.

Of course, $w$ can be 9.89, too. That is, $w$ can be any number $\geq 1$ and $< 10$.

So we get: $A = 10^{s+t} = 10^t 10^s = w \cdot 10^s$, which is often called the scientific notation for numbers, and thus, the highest place value in the antilog $A$ is $10^s$. Still not quite sure?

Suppose for instance, $w = 1.5839$, and $s = 3$.

Then, $A = 1.5839 \cdot 10^3 = 1583.9$, so the highest place value in $A$ is 1000, which is $10^3$.

Suppose for another, $w = 1.5839$, and $s = 0$.

Then, $A = 1.5839 \cdot 10^0 = 1.5839$, so the highest place value in $A$ is 1, which is $10^0$.

Suppose for one more, $w = 3.89$, and $s = -3$. Then, $A = 3.89 \cdot 10^{-3} = 3.89 \cdot 0.001 = 0.00389$.

So the highest place value in $A$ is 0.001, which is $10^{-3}$.

Therefore:

If the index $s \geq 0$, then the antilog $A$ has $s + 1$ digits above the decimal point.

If the index $s < 0$, then the first nonzero appears in the $s^{\text{th}}$ digit below the decimal point.

**In short:**

By the definition for logs, $\log_b A = s + t \Leftrightarrow A = b^{s+t}$, where $s$ is an integer, and $0 \leq t < 1$.

Since $b = 10$, we can set: $\log A = s + t \Leftrightarrow A = 10^{s+t}$.

Thus, we have: $A = 10^{s+t} = 10^t 10^s$.   So we get:

$0 \leq t < 1 \Rightarrow 10^0 \leq 10^t < 10^1 \Rightarrow 1 \leq 10^t < 10 \Rightarrow 10^s \leq 10^t 10^s < 10^{s+1} \Rightarrow 10^s \leq A < 10^{s+1}$.

Therefore, the highest place value in $A$ is $10^s$.

## Suggestions or Solutions
## To the Problem 1 in the Example 5

Suppose that $A$ and $b > 0$, but $b \neq 1$, and that $\log_b A = s + t$, where $s$ is an integer called the index, and $0 \leq t < 1$, where $t$ is called the mantissa.

Then, assuming that $X$ and $Y > 0$, and that $\log X$ and $\log Y$ have the same mantissa, but have different indices, show that both $X$ and $Y$ have the same sequence of digits, but their positions of the decimal points are different.

Since both mantissas are the same, we can set:

$\log X = s + m$, and $\log Y = u + m$, where $s$ and $u$ are integers, and $0 \leq m < 1$.

Then, by the definition for logs, we get:

$\log X = s + m \Leftrightarrow X = 10^{s+m}$, and $\log Y = u + m \Leftrightarrow Y = 10^{u+m}$.

Next, setting $w = 10^m$, we get:

$X = 10^{s+m} = 10^m 10^s = w \cdot 10^s$, and $Y = 10^{u+m} = 10^m 10^u = w \cdot 10^u$.

Next, we have: $0 \leq m < 1$, and $w = 10^m$. So we get:

$10^0 \leq 10^m < 10^1 \Rightarrow 1 \leq 10^m < 10 \Rightarrow 1 \leq w < 10$.

Thus, we get: $X = w \cdot 10^s$, $Y = w \cdot 10^u$, and $1 \leq w < 10$.

Now, $w$ shows the sequence of digits in each of the antilogs $X$ and $Y$.

Also, $u$ and $s$ can respectively indicate the positions of the decimal points in the antilogs $X$ and $Y$.

Therefore, $X$ and $Y$ both have the same sequence of digits, but their positions of the decimal points are different.

*If not quite sure of the idea behind the processes above, follow the steps below:*

Suppose first, that:

$\log X = s + t$, where $s$ is an integer, and $0 \le t < 1$.

$\log Y = u + v$, where $u$ is an integer, and $0 \le v < 1$.

Then, $s$ and $u$ are indices, $t$ and $v$ are mantissas, and by the definition for logs, we get:

$\log X = s + t \Leftrightarrow X = 10^{s+t}$, and $\log Y = u + v \Leftrightarrow Y = 10^{u+v}$.

Next, since the mantissas are to be the same, we can set: $k = t = v$.

Then, we get: $X = 10^{s+k}$, and $Y = 10^{u+k}$, and thus, $X = 10^k 10^s$, and $Y = 10^k 10^u$.

Now, we have: $0 \le t < 1$, and $0 \le v < 1$.  Thus, we get: $0 \le k < 1$ since $k = t = w$.

So we get: $0 \le k < 1 \Rightarrow 10^0 \le 10^k < 10^1 \Rightarrow 1 \le 10^k < 10$.

Suppose next, that $w = 10^k$.  Then, we get: $1 \le w < 10$, and thus, we get:

$X = 10^k 10^s = w \cdot 10^s$, and $Y = 10^k 10^u = w \cdot 10^u$.

Now, $w$ can be any number greater than or equal to 1 and less than 10 such as 3.89053, so $w$ shows the sequence of digits in each of the two antilogs $X$ and $Y$.

That is, $X$ and $Y$ both have the same sequence of digits.

Next, the index $u$ indicates the position of the decimal point in the antilog $X$.

Also, the other index $s$ indicates the position of the decimal point in the antilog $Y$.

For instance, if $s = -2$, and $u = 3$, we get: $X = 0.0389053$, and $Y = 3890.53$.

Therefore, $X$ and $Y$ both have the same sequence of digits, but their positions of the decimal points are different.

## Suggestions or Solutions
## To the Problem 2 in the Example 5

**Suppose that $A$ and $b > 0$, but $b \neq 1$, and that $\log_b A = s + t$, where $s$ is an integer called the index, and $0 \leq t < 1$, where $t$ is called the mantissa.**

**Then, assuming that $\log X - \log Y = k$ where $k$ is an integer, find the ratio, $\frac{X}{Y}$.**

$\log X - \log Y = \log \frac{X}{Y} = k \Leftrightarrow \frac{Y}{X} = 10^k$ by the definition for logs.

Therefore, the ratio $r = \frac{Y}{X} = 10^k$.

*If not quite sure of the idea behind the processes above, follow the steps below:*

We have a log identity, where $\log_b A - \log_b C = \log_b \frac{A}{C}$.
And taking a common log, we usually omit the base 10.

Then, we get: $\log X - \log Y = \log \frac{X}{Y}$.

And we have: $\log X - \log Y = k$, too.

So we get: $\log \frac{X}{Y} = k$.

Then, by the definition for logs, we can get: $\log \frac{X}{Y} = k \Leftrightarrow \frac{Y}{X} = 10^k$.

Therefore, the ratio $\frac{X}{Y} = 10^k$.

So what else can we mean by the fact that $\log X - \log Y = k$ where $k$ is an integer?

Both $\log X$ and $\log Y$ have the same mantissa.

That is, for instance, $\log X = 3.1234$, and $\log X = 2.1234$.

So both antilogs $X$ and $Y$ have the same sequence of digits.

That is, for instance, $X = \mathbf{123.45}$, and $Y = \mathbf{12.345}$.

**In short:**

$\log X - \log Y = \log \frac{X}{Y} = k \Leftrightarrow \frac{Y}{X} = 10^k$ by the definition for logs.

Therefore, the ratio $r = \frac{Y}{X} = 10^k$.

## Examples 4 on Logarithms

In this set of examples, too, we are going to do some practice on basic algebra on logs so that we get more familiar with manipulation of logs. That's because algebra matters.

0.    Find the number of digits in the octal equivalent of a decimal $3^{100}$.
Use **log 2 = 0.3010** and **log 3 = 0.4771**.

1.    A radioisotope Ra has a half-life of 1602 years. A half-life is the time a radioisotope takes to lose half its radioactivity through decay. Assuming the radioactivity of Ra is 8% of the original in an ancient tool, find the age of the tool. Use **log 2 = 0.3010**.

2.    Suppose on education, we will spend 12.7% of the income to be made this year, and income increases by 15.7% each year. Then, by what percent do we have to increase education budget each year during the next 7 years if we want the budget to be 21% of the income of the year after the next seven years? A log table may be used.

## Suggestions or Solutions
## To the Problem in the Example 0

**Find the number of digits in the octal equivalent of a decimal $3^{100}$.**
**Use log 2 = 0.3010 and log 3 = 0.4771.**

$$\log_8 3^{100} = 100\log_8 3 = 100\frac{\log 3}{\log 8} = 100\frac{\log 3}{3\log 2} = \frac{100}{3}\frac{\log 3}{\log 2} = \frac{100}{3}\cdot\frac{0.4771}{0.3010} = \frac{47.71}{0.9030} = \frac{47710}{903} = 52 + \frac{754}{903}.$$

Therefore, the highest place value in the octal equivalent is $8^{52}$, so the octal equivalent has 53 digits. In other words, it is a 53-digit octal integer.

*If not quite sure of the idea behind the processes above, follow the steps below:*

We are given a number decimal, and want to get the number of all the digits in the octal equivalent, which is a number to the base 8. So do we need to convert first, the decimal into the octal equivalent?

No, we don't want to do that. Why not?

If we do so, it will probably take a lot of time, since the decimal given is a huge number.

Using a calculator, we can get $5.1537752073201133103646112976562\cdot10^{47}$, which is in fact, a 48-digit number, which is a humongous number we can even hardly read.

So instead, we may want to use logarithms getting the number of digits.

The number given is $3^{100}$, which is an integer, which is called a number to base **10**, often simply called a decimal number or just called a decimal, for short.

Since the number given is an integer, the octal equivalent is an integer, too. And the number of digits in an integer has much to do with the highest place value in the integer.

The highest place value in a decimal can be indicated by an integer $c$, usually called the characteristic of the common log of the decimal, and in fact, the highest place value in a decimal is $10^c$, where $c$ is the integer called the characteristic.

Taking a common log of $A$ decimal, we can set: $\log A = c + m$, where $c$ is an integer called the characteristic, $m$ is called the mantissa, and $0 \le m < 1$.

The decimal $A$ is an antilog in a log, and thus, is supposed to be positive. If negative, take the absolute value of it, or multiply it by $-1$, and then, take the common log of it.

Suppose now, we want to find the highest place value in a number $B$, which is not a decimal but a number to a base $b$, and $A$ is the decimal equivalent of $B$.

(If for instance, $b$ is 2, $B$ is called <u>a binary number</u>, or just quickly called <u>a binary</u>.)

Then, we can find the highest place value in $B$ taking the log of $A$ to base $b$. Then, we put it this way: $\log_b A = s + t$, where $s$ is an integer, and $0 \le t < 1$. So $t$ is the mantissa. In this case however, $s$ is called the index instead of the characteristic.

Then, $b^s$ is not the highest place value in the decimal $A$ but the highest place value in $B$, which is the number to the base $b$.

(If it sounds however, totally new to you, refer to the section **Number Systems and Logs** in the book, **Powers and Logarithms 2**.)

Solving thus, the problem in this example, we begin with taking a $\log_8$ of the integer $3^{100}$, and then, get the index.    So taking first, the log of it, we get:

$$\log_8 3^{100} = 100 \log_8 3 = 100\,\frac{\log 3}{\log 8} = 100\,\frac{\log 3}{3\log 2} = \frac{100}{3}\,\frac{\log 3}{\log 2} = \frac{100}{3} \cdot \frac{0.4771}{0.3010} = \frac{47.71}{0.9030} = \frac{47710}{903} = 52 + \frac{754}{903}.$$

So the index is **52**, and the mantissa is $\frac{754}{903}$.

Thus, we can see that the highest place value in the octal equivalent is $8^{52}$.

Therefore, the octal equivalent has 53 digits.    How come?

Expanding a decimal 189, we get: $189 = 1 \cdot 10^2 + 8 \cdot 10^1 + 9 \cdot 10^0$.

Thus, 189 in decimal has 3 digits, and the highest place value in 189 is $10^2$.

On the other hand, converting the decimal 189 into the octal equivalent, we get:

$189 = 2 \cdot 8^2 + 7 \cdot 8^1 + 5 \cdot 8^0 \Rightarrow 275$ is the octal equivalent of the decimal 189.

Thus, 275 in octal has 3 digits, and the highest place value in 275 octal is $8^2$.

Let's see now, if taking a $\log_8$ of 189 works.

$$\log_8 189 = \log_8 64 \cdot \frac{189}{64} = \log_8 64 + \log_8 \frac{189}{64} = 2 + \log_8 \frac{189}{64}.$$

We know: $1 \le \frac{189}{64} < 8 \Rightarrow \log_8 1 \le \log_8 \frac{189}{64} < \log_8 8 \Rightarrow 0 \le \log_8 \frac{189}{64} < 1$.

So we can set: $\log_8 189 = 2 + t$, where $0 \le t < 1$.

Therefore, the index is 2, and thus, the highest place value in the octal equivalent to 189 decimal is $8^2$, and the octal equivalent has 3 digits.

Meanwhile, taking a common log of $3^{100}$, we get:

$\log 3^{100} = 100 \log 3 = 100 \cdot 0.4771 = 47.71$.

So the characteristic is 47, which is telling us that the highest place value in $3^{100}$ is $10^{47}$, and that $3^{100}$ has 48 digits. In other words, $3^{100}$ is a 48-digit integer.

**In short:**

$$\log_8 3^{100} = 100 \log_8 3 = 100 \frac{\log 3}{\log 8} = 100 \frac{\log 3}{3 \log 2} = \frac{100}{3} \frac{\log 3}{\log 2} = \frac{100}{3} \cdot \frac{0.4771}{0.3010} = \frac{47.71}{0.9030} = \frac{47710}{903} = 52 + \frac{754}{903}.$$

Therefore, the highest place value in the octal equivalent is $8^{52}$, so the octal equivalent has 53 digits. In other words, it is a 53-digit octal integer.

## Suggestions or Solutions
## To the Problem in the Example 1

**A radioisotope Ra has a half-life of 1602 years. A half-life is the time a radioisotope takes to lose half its radioactivity through decay. Assuming the radioactivity of Ra is 8% of the original in an ancient tool, find the age of the tool. Use log 2 = 0.3010.**

Suppose $A$ is the original radioactivity of **Ra**.

Then, after the $n^{th}$ half-life, the radioactivity is $\frac{A}{2^n}$, and the one at present is $0.08A$.

Thus, $\frac{A}{2^n} = 0.08A \Rightarrow 2^{-n} = 0.08 \Rightarrow 2^{-(n+3)} = 0.01$.

Then, $\log 2^{-(n+3)} = \log 0.01 = -2 \Rightarrow -(n+3)\log 2 = -2 \Rightarrow n+3 = \frac{2}{\log 2} \Rightarrow n = \frac{2}{\log 2} - 3$.

$\log 2 = 0.3010 \Rightarrow n = \frac{2}{\log 2} - 3 = \frac{2}{0.3010} - 3 = \frac{2-0.9030}{0.3010} = \frac{1.097}{0.3010} = \frac{1097}{301}$.

Therefore, the age of the tool is $1602n = 1602 \cdot \frac{1097}{301} \cong 5838.52$ years old.

*If not quite sure of the idea behind the processes above, follow the steps below:*

The solution to this problem is the age of the tool, of course.
Basically though, what do we want to get first, finding the solution?

It is the number of half-lives.
More specifically, it is the number of those taken until the radioactivity of **Ra** decrease to 8% of the original.   How can we find the number, then?

We can try keeping track of **Ra**'s decay over some half-lives.
What then, do we want to begin with?

We may want to begin with setting up the initial and the final states of **Ra**.

So let's suppose $A$ is the original radioactivity.

Then, the final, that is, the present radioactivity is 8% of $A$, which is **0.08$A$**.

Now, let's keep track of **Ra**'s decay over several half-lives.

After the first half-life, the radioactivity is half the original $A$, so we get: $\frac{A}{2}$. $\{A/2\}$

After the second one, it is a half of the half, so we get: $\frac{\frac{A}{2}}{2} = \frac{A}{2^2}$. $\{(A/2)/2 = A/2/2 = A/2^2\}$

After the third, it's half the amount above, so we get: $\frac{\frac{A}{2^2}}{2} = \frac{A}{2^3}$. $\{(A/2^2)/2 = A/2/2/2 = A/2^3\}$

Thus, after the $n^{\text{th}}$ one, we get: $\frac{A}{2^n}$. $\{A/2^n\}$

(In fact, $n$ doesn't have to be an integer. We will see why not shortly.)

The radioactivity at present is **0.08$A$**.  So we can set: $\frac{A}{2^n} = 0.08A$, and thus, we get:

$$\frac{A}{2^n} = 0.08A \Rightarrow \frac{1}{2^n} = 0.08 \Rightarrow 2^{-n} = 0.08 = 0.01 \cdot 8 \Rightarrow \frac{2^{-n}}{8} = 0.01 \Rightarrow 2^{-(n+3)} = 0.01.$$

Now, we want to solve for $n$ the equation above, which is an exponential equation
Solving such an equation, what do we get as the solution?

We get the exponent, which is another name for a log.

So since it's an exponential equation, we may want to take logs of both sides.
What log then, do we want to take?

We are given **log 2 = 0.3010**, which is a common log. Thus, common logs look the best.

So taking the log of each side, we get: $\log 2^{-(n+3)} = \log 0.01$.

Then, beginning with the right hand sides, we get: $\log 0.01 = \log 10^{-2} = -2$.

Next, moving on to the left hand side, we get: $\log 2^{-(n+3)} = -(n+3)\log 2$.

So we get: $\log 2^{-(n+3)} = \log 0.01 \Rightarrow -(n+3)\log 2 = -2 \Rightarrow n+3 = \frac{2}{\log 2} \Rightarrow n = \frac{2}{\log 2} - 3$.

And we have: $\log 2 = 0.3010$, too.

So we get: $n \cong \frac{2}{\log 2} - 3 = \frac{2}{0.3010} - 3 = \frac{2-0.9030}{0.3010} = \frac{1.097}{0.3010} = \frac{1097}{301}$.

Now, $n$ was thought to be the number of half-lives, but doesn't look like an integer.

It does not have to be an integer. In fact, $n \cong \frac{1097}{301} = 3.644518...$, so it isn't an integer.

Why not, and why not = sign but $\cong$ sign?

Usually, values from a log table are not exact but approximate. And we have:

$n \cong \frac{1097}{301} = \frac{903+194}{301} = 3 + \frac{194}{301}$, so after the third half-life, it only takes $\frac{194}{301}$ of the fourth one to reach 8% of the original. Still not quite sure?

When keeping track of **Ra**'s decay, we've had:

After the first, $\frac{4}{2}$, which is $A/2$. After the second, $\frac{\frac{4}{2}}{2} = \frac{4}{2^2}$, which is $(A/2)/2 = A/2/2 = A/2^2$.

After the third, $\frac{\frac{4}{2^2}}{2} = \frac{4}{2^3}$, which is $(A/2^2)/2 = A/2/2/2 = A/2^{1+1+1} = A/2^3$.

Now, $in$ the fourth, $(A/2^3)/2^{\frac{194}{301}} = A/2/2/2/2^{\frac{194}{301}} = A/2^{1+1+1+\frac{194}{301}} = A/2^{3+\frac{194}{301}} = \dfrac{A}{2^{\frac{1097}{301}}}$.

We know that **Ra** has a half-life of 1602 years.   So one full half-life takes 1602 years.

Thus, the age of the tool is $1602n \cong 1602 \cdot \frac{1097}{301} \cong 5838.52$ years old.

**In short:**

Suppose $A$ is the original radioactivity of **Ra**.

Then, after the $n^{th}$ half-life, the radioactivity is $\frac{A}{2^n}$, and the one at present is $0.08A$.

Thus, $\frac{A}{2^n} = 0.08A \Rightarrow 2^{-n} = 0.08 \Rightarrow 2^{-(n+3)} = 0.01$.

Then, $\log 2^{-(n+3)} = \log 0.01 = -2 \Rightarrow -(n+3) \log 2 = -2 \Rightarrow n+3 = \frac{2}{\log 2} \Rightarrow n = \frac{2}{\log 2} - 3$.

$\log 2 = 0.3010 \Rightarrow n = \frac{2}{\log 2} - 3 = \frac{2}{0.3010} - 3 = \frac{2 - 0.9030}{0.3010} = \frac{1.097}{0.3010} = \frac{1097}{301}$.

Therefore, the age of the tool is $1602n = 1602 \cdot \frac{1097}{301} \cong 5838.52$ years old.

## Suggestions or Solutions
## To the Problem in the Example 2

**Suppose on education, we will spend 12.7% of the income to be made this year, and income increases by 15.7% each year.**
**Then, by what percent do we have to increase education budget each year during the next 7 years if we want the budget to be 21% of the income of the year after the next seven years? A log table may be used.**

Suppose that $R_n$ is the annual income, and that $E_n$ is the annual spending on education.

Suppose also, $n = 0$ for this year. Then, we get:

$E_0 = \frac{12.7}{100} R_0 = 0.127R_0 \Rightarrow E_0 = 0.127R_0.$

$R_1 = R_0 + \frac{15.7}{100} R_0 = R_0 + 0.157R_0 = R_0(1 + 0.157).$

$R_2 = R_1 + 15.7\% \text{ of } R_1 = R_1 + \frac{15.7}{100} R_1 = R_1 + 0.157R_1 = R_1(1 + 0.157)$
$= R_0(1 + 0.157)^2 = 1.157^2 R_0.$

So we get: $R_{n-1} = 1.157^{(n-1)} R_0$. Thus, the income in the $8^{\text{th}}$ year is $R_7 = 1.157^7 R_0$.

Suppose now, that $A\%$ is the annual increase on education budget. Then:

$E_1 = E_0 + A\% \text{ of } E_0 = E_0 + \frac{A}{100} E_0 = E_0(1 + 0.01A).$

$E_2 = E_1 + A\% \text{ of } E_1 = E_1 + \frac{A}{100} E_1 = E_1(1 + 0.01A) = E_0(1 + 0.01A)^2.$ Set: $B = 0.01A$.

Then, we get: $E_{n-1} = E_0(1 + B)^{n-1}.$   Thus, the budget in the $8^{\text{th}}$ year is $E_7 = E_0(1 + B)^7.$

Since $E_7$ is 21% of $R_7$, we get: $E_7 = 0.21R_7 \Rightarrow E_0(1 + B)^7 = (0.21)1.157^7 R_0.$

We have: $E_0 = 0.127R_0$, too.

Thus, $E_0(1 + B)^7 = (0.21)1.157^7 R_0 \Rightarrow 0.127 R_0 (1 + B)^7 = (0.21)1.157^7 R_0$

$\Rightarrow 0.127(1 + B)^7 = (0.21)1.157^7 \Rightarrow 1.27(1 + B)^7 = (2.1)1.157^7.$

Taking logs of both sides, we get: $\log 1.27(1 + B)^7 = \log (2.1)1.157^7.$

$\log 1.27 + \log (1 + B)^7 = \log 2.1 + \log 1.157^7$
$\Rightarrow \log (1 + B)^7 = \log 2.1 + \log 1.157^7 - \log 1.27$
$\Rightarrow 7\log (1 + B) = \log 2.1 + 7\log 1.157 - \log 1.27$
$\Rightarrow \log (1 + B) = \log 1.157 + \frac{\log 2.1 - \log 1.27}{7}.$

Using a log table, we get: $\log 1.157 = 0.0634$, $\log 2.1 = 0.3222$, and $\log 1.27 = 0.1038$.

Thus, we get: $\log (1 + B) = \log 1.157 + \frac{\log 2.1 - \log 1.27}{7} \cong 0.0634 + \frac{0.3222 - 0.1038}{7}$
$= 0.0634 + 0.0312 = 0.0946 \Rightarrow \log (1 + B) \cong 0.0946.$

Using a log table again, we get: $1 + B \cong 1.24 \Rightarrow B \cong 0.24.$

We have set: $B = 0.01A$ earlier, so we get: $0.01A \cong 0.24 \Rightarrow A \cong 24.$

Therefore, we want to increase the education budget by approximately 24% each year during the next 7 years.

*If not quite sure of the idea behind the processes above, follow the steps below:*

This example is basically the same as the example 7. The solution to this is though, not the number of half-lives but the amount of a half-life if you will. That is, it is not the number of years but the percentage increase. So in this case, too, we may want to begin with setting the initial and final states of the object under consideration. Unlike the case in the example 7 though, the number of such objects is not one but two. One is income, and the other is the budget, which is education spending.   What then, is the next?

We can try keeping track of each object.

So first, assuming $R_0$ is the income this year, and $E_0$ is the spending on education this year, we get: $E_0 = 12.7\%$ of $R_0 = \frac{12.7}{100} R_0 = 0.127R_0 \Rightarrow E_0 = 0.127R_0$.

Next, the annual income increases by 15.7% during the next 7 years. So:

The income in this year, that is, the first year $= R_0$.

The income in the second year

$= R_1 = R_0 + 15.7\%$ of $R_0 = R_0 + \frac{15.7}{100} \cdot R_0 = R_0 + 0.157R_0 = R_0(1 + 0.157)$.

The income in the third year

$= R_2 = R_1 + 15.7\%$ of $R_1 = R_1 + \frac{15.7}{100} \cdot R_1 = R_1 + 0.157R_1 = R_1(1 + 0.157)$.

We have: $R_1 = R_0(1 + 0.157) = 1.157R_0$, too.

So we get: $R_2 = R_1(1 + 0.157) = 1.157R_1 = 1.157 \cdot 1.157R_0 = 1.157^2 R_0$.

Thus, by the same token, the income in the $n^{\text{th}}$ year, we get: $R_{n-1} = 1.157^{(n-1)} R_0$.

Now, the year after the next seven years is the $8^{\text{th}}$ year.

So the income in the $8^{\text{th}}$ year is $R_7 = 1.157^7 R_0$.

Suppose now, that $A\%$ is the annual increase on education budget.

We know $E_0$ is the education budget this year. So:

The budget in the second year is $E_1 = E_0 + A\%$ of $E_0 = E_0 + \frac{A}{100} E_0 = E_0(1 + 0.01A)$.

The budget in the third year is

$E_2 = E_1 + A\%$ of $E_1 = E_1 + \frac{A}{100} E_1 = E_1(1 + 0.01A) = E_0(1 + 0.01A)(1 + 0.01A)$.

Thus, the budget in the third year is $E_2 = E_0(1 + 0.01A)^2$.

By the same token, in the $n^{th}$ year, setting $B = 0.01A$, we get: $E_{n-1} = E_0(1 + B)^{n-1}$.

Now, the year after the next seven years is the 8$^{th}$ year.

So the budget in the 8$^{th}$ year is $E_7 = E_0(1 + B)^7$, where $E_0 = 0.127R_0$, and $B = 0.01A$.

Now, we want $E_7$ to be 21% of $R_7$, which is the income in the 8$^{th}$ year, and is $1.157^7R_0$.

So we get: $E_7 = \frac{21}{100}R_7 = 0.21R_7 \Rightarrow E_7 = 0.21R_7 \Rightarrow E_0(1 + B)^7 = 0.21 \cdot 1.157^7R_0$.

And we have $E_0 = 0.127R_0$, too.

So we get: $E_0(1 + B)^7 = 0.21 \cdot 1.157^7R_0 \Rightarrow 0.127R_0(1 + B)^7 = 0.21 \cdot 1.157^7R_0$.

Canceling out $R_0$, we get: $0.127(1 + B)^7 = 0.21 \cdot 1.157^7$.

Let's put it this way, though. Multiplying both sides by 10, we get:

$1.27(1 + B)^7 = 2.1 \cdot 1.157^7$, because we are going to use a log table.

Now, we get to solve the above for $B$.
So we want to remove the exponent 7 from $(1 + B)^7$. Exponent is another name for a log.

So taking logs of both sides, we get: $\log 1.27(1 + B)^7 = \log 2.1 \cdot 1.157^7$, and we get:

$\log 1.27 + \log (1 + B)^7 = \log 2.1 + \log 1.157^7$

$\Rightarrow \log (1 + B)^7 = \log 2.1 + \log 1.157^7 - \log 1.27$

$\Rightarrow 7 \log (1 + B) = \log 2.1 + 7 \log 1.157 - \log 1.27$

$\Rightarrow \log (1 + B) = \log 1.157 + \dfrac{\log 2.1 - \log 1.27}{7}$.

Let's now use a log table.

Then, we get: **log 1.157 = 0.0634**, **log 2.1 = 0.3222**, and **log 1.27 = 0.1038**.

Thus, we get: $\log (1 + B) = \log 1.157 + \frac{\log 2.1 - \log 1.27}{7} \cong 0.0634 + \frac{0.3222 - 0.1038}{7}$

$= 0.0634 + 0.0312 = 0.0946 \Rightarrow \log (1 + B) \cong 0.0946$.

Usually, values from a log table are not exact but approximate, so we want to use $\cong$ sign and not = sign.

Now, using a log table again, we get $1 + B \cong 1.24$.

Thus, we get: $B \cong 0.24$. We have set: $B = 0.01A$, earlier.

So we get: $0.01A \cong 0.24 \Rightarrow A \cong 24$.

Therefore, we want to increase the education budget by approximately 24% each year during the next 7 years.

Using a calculator, we get: $\log (1 + B) \cong 0.0946 \Rightarrow 1 + B \cong 10^{0.0946} \cong 1.2434$.

**In short:**

Suppose that $R_n$ is the annual income, and that $E_n$ is the annual spending on education.

Suppose also, $n = 0$ for this year. Then, we get:

$E_0 = \frac{12.7}{100} R_0 = 0.127R_0 \Rightarrow E_0 = 0.127R_0$.

$R_1 = R_0 + \frac{15.7}{100} R_0 = R_0 + 0.157R_0 = R_0(1 + 0.157)$.

$R_2 = R_1 + 15.7\% \text{ of } R_1 = R_1 + \frac{15.7}{100} R_1 = R_1 + 0.157R_1 = R_1(1 + 0.157)$
$= R_0(1 + 0.157)^2 = 1.157^2R_0$.

So we get: $R_{n-1} = 1.157^{(n-1)}R_0$. Thus, the income in the 8th year is $R_7 = 1.157^7R_0$.

Suppose now, that $A\%$ is the annual increase on education budget. Then:

$E_1 = E_0 + A\%$ of $E_0 = E_0 + \frac{A}{100}E_0 = E_0(1 + 0.01A)$.

$E_2 = E_1 + A\%$ of $E_1 = E_1 + \frac{A}{100}E_1 = E_1(1 + 0.01A) = E_0(1 + 0.01A)^2$. Set: $B = 0.01A$.

Then, we get: $E_{n-1} = E_0(1 + B)^{n-1}$.   Thus, the budget in the $8^{th}$ year is $E_7 = E_0(1 + B)^7$.

Since $E_7$ is 21% of $R_7$, we get: $E_7 = 0.21R_7 \Rightarrow E_0(1 + B)^7 = (0.21)1.157^7R_0$.

We have: $E_0 = 0.127R_0$, too.

Thus, $E_0(1 + B)^7 = (0.21)1.157^7R_0 \Rightarrow 0.127R_0(1 + B)^7 = (0.21)1.157^7R_0$

$\Rightarrow 0.127(1 + B)^7 = (0.21)1.157^7 \Rightarrow 1.27(1 + B)^7 = (2.1)1.157^7$.

Taking logs of both sides, we get: $\log 1.27(1 + B)^7 = \log (2.1)1.157^7$.

$\log 1.27 + \log (1 + B)^7 = \log 2.1 + \log 1.157^7$

$\Rightarrow \log (1 + B)^7 = \log 2.1 + \log 1.157^7 - \log 1.27$

$\Rightarrow 7\log (1 + B) = \log 2.1 + 7\log 1.157 - \log 1.27$

$\Rightarrow \log (1 + B) = \log 1.157 + \frac{\log 2.1 - \log 1.27}{7}$.

Using a log table, we get: $\log 1.157 = 0.0634$, $\log 2.1 = 0.3222$, and $\log 1.27 = 0.1038$.

Thus, we get: $\log (1 + B) = \log 1.157 + \frac{\log 2.1 - \log 1.27}{7} \cong 0.0634 + \frac{0.3222 - 0.1038}{7}$

$= 0.0634 + 0.0312 = 0.0946 \Rightarrow \log (1 + B) \cong 0.0946$.

Using a log table again, we get: $1 + B \cong 1.24 \Rightarrow B \cong 0.24$.

We have set: $B = 0.01A$ earlier, so we get: $0.01A \cong 0.24 \Rightarrow A \cong 24$.

Therefore, we want to increase the education budget by approximately 24% each year during the next 7 years.

## Examples 5 in Logarithms

In this set of examples, too, we are going to do some practice on basic algebra on logs so that we get used to manipulation of logs. That's because, of course, algebra matters.

0.0.   Assuming $\log_{16} 7 = x$ and $16^y = 17$, put $\log_{64} 119$ in terms of $x$ and $y$.

0.1.   Assuming $\log_{14} 7 = x$ and $14^y = 8$, put $\log_{28} 56$ in terms of $x$ and $y$.

1.   Show that $(\log_{15} 3)^2 + \dfrac{\log_{15} 45}{\log_5 15} = 1$.

2.   Suppose that $a$, $b$, and $c > 0$, but that $a$, $b$, and $c \neq 1$. Suppose also, that:

$p$, $q$, and $r \neq 0$.

$pq + qr + rp = pqr$.

$\log_a t = p$, $\log_b t = q$, and $\log_c t = r$.

Then, put $t$ in terms of $a$, $b$, and $c$.

3.   Assuming $1 < c < d$, and $c^{\frac{1}{m}} = d^{\frac{1}{n}} = cd$, show that: $\sqrt{m^3 - m^2 n - mn^2 + n^3} = \log_{dc} \frac{d}{c}$.

## Suggestions or Solutions
## To the Problem 0 in the Example 0

**Assuming $\log_{16} 7 = x$, and $16^y = 17$, put $\log_{64} 119$ in terms of $x$ and $y$.**

$$16^y = 17 \Rightarrow \log 16^y = \log 17 \Rightarrow y \log 16 = \log 17 \Rightarrow y = \frac{\log 17}{\log 16} = \log_{16} 17.$$

$$\log_{64} 119 = \frac{\log_{16} 119}{\log_{16} 64}. \quad \log_{16} 64 = \log_{4^2} 4^3 = \tfrac{3}{2} \log_4 4 = \tfrac{3}{2}. \quad \log_{16} 119 = \log_{16} 17 \cdot 7 = x + y.$$

Therefore, $\log_{64} 119 = \frac{x+y}{\frac{3}{2}} = \frac{2}{3}(x + y)$.

*If not quite sure of the idea behind the processes above, follow the steps below:*

Since we need to put something in terms of $x$ and $y$, we want to know what $x$ and $y$ are. We have: $x = \log_{16} 7$. What about $y$?

We have: $16^y = 17$, which is an exponential equation, so we want to solve it for $y$.

Solving such an equation as above, we get as the solution an exponent, which is another name for a log. So taking logs of both sides, we get:

$\log 16^y = \log 17 \Rightarrow y \log 16 = \log 17 = \dfrac{\log 17}{\log 16}$. Besides, we have: $\log_a b = \dfrac{\log_c b}{\log_c a}$, too.

So we get: $y = \dfrac{\log 17}{\log 16} = \log_{16} 17$.

Thus, we now have: $x = \log_{16} 7$, and $y = \log_{16} 17$, both of which are to base 16.

So we want to put $\log_{64} 119$ in some logs to base 16.　How?

Using the same identity as above again, we get: $\log_{64} 119 = \dfrac{\log_{16} 119}{\log_{16} 64}$.

Meanwhile, $\log_{16} 64 = \log_{16} 16 \cdot 4 = \log_{16} 16 + \log_{16} 4 = 1 + \log_{16} 16^{0.5} = 1 + 0.5 = 1.5$.

Thus, we get: $\log_{64} 119 = \dfrac{\log_{16} 119}{1.5} = \dfrac{\log_{16} 119}{\frac{3}{2}} = \frac{2}{3} \log_{16} 119$.  Now, we should be able to put 119 in terms of 7 and 17 since we have: $x = \log_{16} 7$, and $y = \log_{16} 17$. Factorizing 119, we get: $119 = 17 \cdot 7$.

So we get: $\log_{16} 119 = \log_{16} 7 \cdot 17 = \log_{16} 7 + \log_{16} 17$, which is $x + y$.

Therefore, $\log_{64} 119 = \frac{2}{3}(x + y)$.

**In short:**

$16^y = 17 \Rightarrow \log 16^y = \log 17 \Rightarrow y \log 16 = \log 17 \Rightarrow y = \dfrac{\log 17}{\log 16} = \log_{16} 17$.

$\log_{64} 119 = \dfrac{\log_{16} 119}{\log_{16} 64}$.   $\log_{16} 64 = \log_{4^2} 4^3 = \frac{3}{2} \log_4 4 = \frac{3}{2}$.   $\log_{16} 119 = \log_{16} 17 \cdot 7 = x + y$.

Therefore, $\log_{64} 119 = \dfrac{x+y}{\frac{3}{2}} = \frac{2}{3}(x + y)$.

## Suggestions or Solutions
## To the Problem 1 in the Example 0

**Assuming $\log_{14} 7 = x$, and $14^y = 8$, put $\log_{28} 56$ in terms of $x$ and $y$.**

To begin with, $\log 14^y = \log 8 \Rightarrow y \log 14 = \log 8 \Rightarrow y = \dfrac{\log 8}{\log 14} = \log_{14} 8$.

Next, $\log_{28} 56 = \dfrac{\log_{14} 56}{\log_{14} 28} = \dfrac{\log_{14} 7 + \log_{14} 8}{\log_{14} 14 + \log_{14} 2} = \dfrac{x+y}{1 + \log_{14} 2}$.

Meanwhile, $\log_{14} 2 = \log_{14} \frac{14}{2} = \log_{14} 14 - \log_{14} 7 = 1 - x$.

Therefore, $\log_{28} 56 = \dfrac{x+y}{1 + \log_{14} 2} = \dfrac{x+y}{1 + 1 - x} = \dfrac{x+y}{2-x}$.

*If not quite sure of the idea behind the processes above, follow the steps below:*

Putting something in terms of $x$ and $y$, we want to know what $x$ and $y$ are.
We have: $x = \log_{14} 7$, but not $y$. So what about $y$?

We can get $y$ from one of the two equations above, which is $14^y = 8$.
So we want to solve an exponential equation.
Solving such an equation, we often take logs of both sides of the equation.

So taking logs of both sides in the equation $14^y = 8$, we get:

$\log 14^y = \log 8 \Rightarrow y \log 14 = \log 8 \Rightarrow y = \dfrac{\log 8}{\log 14} = \log_{14} 8$.

Thus, we now have: $x = \log_{14} 7$, and $y = \log_{14} 8$, both of which are to base 14, so we want to put $\log_{28} 56$ in terms of logs to base 14.

We have an identity, $\log_a b = \log_a b = \dfrac{\log_c b}{\log_c a}$, so we can set: $\log_{28} 56 = \dfrac{\log_{14} 56}{\log_{14} 28}$.

We know: $x + y = \log_{14} 7 + \log_{14} 8 = \log_{14} 56$, since $56 = 7 \cdot 8$. What then, about $\log_{14} 28$?

We have: $28 = 14 \cdot 2$, so we get: $\log_{14} 28 = 1 + \log_{14} 2$.　 Well then, what about 2?

We have: $2 = \frac{14}{7}$, so we get: $\log_{14} 2 = \log_{14} \frac{14}{7} = 1 - \log_{14} 7 = 1 - x$, since $x = \log_{14} 7$.

Thus, we get: $\log_{14} 28 = 1 + \log_{14} 2 = 1 + (1 - x) = 2 - x$.

Therefore, we can see that $\log_{28} 56 = \dfrac{\log_{14} 56}{\log_{14} 28} = \dfrac{x + y}{2 - x}$.

**In short:**

To begin with, $\log 14^y = \log 8 \Rightarrow y \log 14 = \log 8 \Rightarrow y = \dfrac{\log 8}{\log 14} = \log_{14} 8$.

Next, $\log_{28} 56 = \dfrac{\log_{14} 56}{\log_{14} 28} = \dfrac{\log_{14} 7 + \log_{14} 8}{\log_{14} 14 + \log_{14} 2} = \dfrac{x + y}{1 + \log_{14} 2}$.

Meanwhile, $\log_{14} 2 = \log_{14} \frac{14}{2} = \log_{14} 14 - \log_{14} 7 = 1 - x$.

Therefore, $\log_{28} 56 = \dfrac{x + y}{1 + \log_{14} 2} = \dfrac{x + y}{1 + 1 - x} = \dfrac{x + y}{2 - x}$.

## Suggestions or Solutions
## To the Problem in the Example 1

Show that $(\log_{15} 3)^2 + \dfrac{\log_{15} 45}{\log_5 15} = 1$.

To begin with, $\dfrac{\log_{15} 45}{\log_5 15} = \log_{15} 45 \cdot \dfrac{1}{\log_5 15}$.

Meanwhile, $\dfrac{1}{\log_5 15} = \dfrac{1}{\frac{\log_{15} 15}{\log_{15} 5}} = \dfrac{1}{\frac{1}{\log_{15} 5}} = (\log_{15} 5)$.

So $\dfrac{\log_{15} 45}{\log_5 15} = \log_{15} 45 \cdot \dfrac{1}{\log_5 15} = (\log_{15} 45)(\log_{15} 5)$.

Next, $(\log_{15} 45)(\log_{15} 5) = (\log_{15} 5 + \log_{15} 9)(\log_{15} 5) = (\log_{15} 5 + 2\log_{15} 3)(\log_{15} 5)$

$= (\log_{15} 5)^2 + 2(\log_{15} 3)(\log_{15} 5)$.

Thus, $(\log_{15} 3)^2 + \dfrac{\log_{15} 45}{\log_5 15}$

$= (\log_{15} 3)^2 + (\log_{15} 5)^2 + 2(\log_{15} 3)(\log_{15} 5) = (\log_{15} 3 + \log_{15} 5)^2 = (\log_{15} 15)^2 = 1$.

*If not quite sure of the idea behind the processes above, follow the steps below:*

Consistency matters in math as well as others. It is particularly so in math.
To begin with, checking the bases in all the logs in the equality above, we can see that
**log₅ 15** only has a different base.
So we may want to make it have the base 15, too.   How?

We can apply a log identity, $\log_a b = \frac{1}{\log_b a}$, where both $a$ and $b > 0$, but $\neq 1$, of course.

Then, we get: $\log_5 15 = \frac{1}{\log_{15} 5}$.

So we can put the equality above this way, too: $(\log_{15} 3)^2 + (\log_{15} 45)(\log_{15} 5) = 1$.

Now, what we are going to do is to force the right side to be 1, so it looks like we've got to do some algebra.

First, the square term, $(\log_{15} 3)^2$ is **not** $\log_{15} 3^2$, so there seems to be not much we can do about it, and thus, we may want to keep it intact, for now.

Then, it looks like we want to do some modifications to $(\log_{15} 45)(\log_{15} 5)$.

So we are going to break it into pieces, and then, put the pieces together. We don't just break it, of course, and we don't just put them together, either. Both have to add up.
Well, that's basically what we do in math.
Thus, doing the modification, we want to keep in mind that the sum of the modification result and $(\log_{15} 3)^2$ should add up to 1.

So let's have a closer look at the right hand side.
Closely looking at all the numbers used, we can notice that they are made of 3 and 5.

$15 = 3 \cdot 5$, and $45$ can be factorized to $9 \cdot 5 = 3^2 \cdot 5$.

We know that the base is 15, and that $\log_{15} 15 = 1$.

Besides, we know: $\log_{15} 3 + \log_{15} 5 = \log_{15} 3 \cdot 5 = \log_{15} 15 = 1$, also.

In addition, we know: $(\log_{15} 3 + \log_{15} 5)^2 = (\log_{15} 15)^2 = 1$, too.

Now, expanding $(\log_{15} 3 + \log_{15} 5)^2$, we get: $(\log_{15} 3)^2 + 2(\log_{15} 3)(\log_{15} 5) + (\log_{15} 5)^2$.

Next, let's have a closer look at the two equalities below:

$(\log_{15} 3)^2 + (\log_{15} 45)(\log_{15} 5) = 1$, which is the equality we want to show.

$(\log_{15} 3)^2 + 2(\log_{15} 3)(\log_{15} 5) + (\log_{15} 5)^2 = 1$, which is the expansion above.

Then, it looks we want to modify $(\log_{15} 45)(\log_{15} 5)$ to $2(\log_{15} 3)(\log_{15} 5) + (\log_{15} 5)^2$.

Quite often though in math, going backward seems to work better.

Thus, we may want to go from $2(\log_{15} 3)(\log_{15} 5) + (\log_{15} 5)^2$ to $(\log_{15} 45)(\log_{15} 5)$, and then, we can readily go back from $(\log_{15} 45)(\log_{15} 5)$ to $2(\log_{15} 3)(\log_{15} 5) + (\log_{15} 5)^2$.

To begin with, we can have: $2(\log_{15} 3) = 2 \log_{15} 3 = \log_{15} 3^2 = \log_{15} 9$.

So we get: $2(\log_{15} 3)(\log_{15} 5) + (\log_{15} 5)^2 = (\log_{15} 9)(\log_{15} 5) + (\log_{15} 5)^2$

$= (\log_{15} 9 + \log_{15} 5)(\log_{15} 5) = (\log_{15} 45)(\log_{15} 5)$.

That's because we can have: $\log_{15} 9 + \log_{15} 5 = \log_{15} 9 \cdot 5 = \log_{15} 45$.

Thus, we get: $2(\log_{15} 3)(\log_{15} 5) + (\log_{15} 5)^2 = (\log_{15} 45)(\log_{15} 5)$.

**In short:**

To begin with, $\dfrac{\log_{15} 45}{\log_5 15} = \log_{15} 45 \cdot \dfrac{1}{\log_5 15}$.

Meanwhile, $\dfrac{1}{\log_5 15} = \dfrac{1}{\frac{\log_{15} 15}{\log_{15} 5}} = \dfrac{1}{\frac{1}{\log_{15} 5}} = (\log_{15} 5)$.

So $\dfrac{\log_{15} 45}{\log_5 15} = \log_{15} 45 \cdot \dfrac{1}{\log_5 15} = (\log_{15} 45)(\log_{15} 5)$.

Next, $(\log_{15} 45)(\log_{15} 5) = (\log_{15} 5 + \log_{15} 9)(\log_{15} 5) = (\log_{15} 5 + 2 \log_{15} 3)(\log_{15} 5)$

$= (\log_{15} 5)^2 + 2(\log_{15} 3)(\log_{15} 5)$.

Thus, $(\log_{15} 3)^2 + \dfrac{\log_{15} 45}{\log_5 15}$

$= (\log_{15} 3)^2 + (\log_{15} 5)^2 + 2(\log_{15} 3)(\log_{15} 5) = (\log_{15} 3 + \log_{15} 5)^2 = (\log_{15} 15)^2 = 1$.

## Suggestions or Solutions
## To the Problem in the Example 2

**Suppose that $a$, $b$, and $c > 0$, but that $a$, $b$, and $c \neq 1$.**

**Suppose also, that:**

**$p$, $q$, and $r \neq 0$.**

**$pq + qr + rp = pqr$.**

**$\log_a t = p$, $\log_b t = q$, and $\log_c t = r$.**

**Then, put $t$ in terms of $a$, $b$, and $c$.**

To begin with, $\log_a t = p \Rightarrow \dfrac{1}{\log_t a} = p \Rightarrow \log_t a = \dfrac{1}{p}$.

$\log_b t = q \Rightarrow \dfrac{1}{\log_t b} = q \Rightarrow \log_t b = \dfrac{1}{q}$.   $\log_c t = r \Rightarrow \dfrac{1}{\log_t c} = r \Rightarrow \log_t c = \dfrac{1}{r}$.

Next, $pq + qr + rp = pqr \Rightarrow \dfrac{1}{r} + \dfrac{1}{p} + \dfrac{1}{q} = 1$.

So we get: $\dfrac{1}{p} + \dfrac{1}{q} + \dfrac{1}{r} = 1 \Rightarrow \log_t a + \log_t b + \log_t c = 1$.

Also, $\log_t a + \log_t b + \log_t c = \log_t abc$, too.

Therefore, $\log_t abc = 1 \Rightarrow abc = t^1 = t \Rightarrow t = abc$.

*If not quite sure of the idea behind the processes above, follow the steps below:*

Examining expressions above, we can notice that $p$, $q$, and $r$ are all in the same situation, because $p$, $q$, and $r$ are connected to each other via the expression: $pq + qr + rp = pqr$.

In fact, they all have the same condition, too, where each of them $\neq 0$.

Also, looking at equalities, $\log_a t = p$, $\log_b t = q$, and $\log_c t = r$, we can notice $a$ and $p$ are correlated via $t$, $b$ and $q$ are correlated via $t$, and so are $c$ and $r$.

Besides, $a$, $b$, and $c$ all have the same condition, too, where they are all $> 0$, and $\neq 1$.

Thus, we can see that $a$, $b$, and $c$ are all in the same situation, too.
So where do we begin, and what do we work with?

Examining again, all the expressions given, we can notice that $t$ is common in the three equalities, and all the other, $a$, $b$, etc. are in the three, too.

So putting $t$ in terms of $a$, $b$, and $c$, we seem to want to begin with the three equalities:

$\log_a t = p$, $\log_b t = q$, and $\log_c t = r$.

However, all logs in the equalities have different bases. In math, consistency matters. Thus, we may want to let them have the same base.

Using a log identity, $\log_x y = \dfrac{1}{\log_y x}$, we can make each have a base $t$.    Then, we get:

$$\log_a t = p \Rightarrow \frac{1}{\log_t a} = p \Rightarrow \log_t a = \frac{1}{p}. \quad \log_b t = q \Rightarrow \frac{1}{\log_t b} = q \Rightarrow \log_t b = \frac{1}{q}.$$

$$\log_c t = r \Rightarrow \frac{1}{\log_t c} = r \Rightarrow \log_t c = \frac{1}{r}. \quad \text{However, we are not given} \frac{1}{p}, \frac{1}{q}, \text{and } \frac{1}{r}.$$

We can come up with those, though. Taking a closer look at the expression, $pq + qr + rp = pqr$, we can see the fractions there.    How?

Examining each term in the expression closely, we can notice that the divisions of both sides by $pqr$ can give us the ones we want.

We can do so, since $pqr \neq 0$, because $p$, $q$, and $r \neq 0$.

So doing such divisions to both of the sides, we get:

$$pq + qr + rp = pqr \Rightarrow \frac{1}{r} + \frac{1}{p} + \frac{1}{q} = 1.$$

Now, via the last expression above, we can put together all the three logs as follows:

$$\log_t a = \frac{1}{p}, \; \log_t b = \frac{1}{q}, \; \text{and} \; \log_t c = \frac{1}{r}.$$

Then, we get: $\dfrac{1}{p} + \dfrac{1}{q} + \dfrac{1}{r} = 1 \Rightarrow \log_t a + \log_t b + \log_t c = 1.$

Besides, we have: $\log_t a + \log_t b + \log_t c = \log_t abc$, too.

So we get: $\log_t abc = 1 \Rightarrow abc = t^1 = t.$　　Therefore, $t = abc$.

**In short:**

To begin with, $\log_a t = p \Rightarrow \dfrac{1}{\log_t a} = p \Rightarrow \log_t a = \dfrac{1}{p}.$

$\log_b t = q \Rightarrow \dfrac{1}{\log_t b} = q \Rightarrow \log_t b = \dfrac{1}{q}. \quad \log_c t = r \Rightarrow \dfrac{1}{\log_t c} = r \Rightarrow \log_t c = \dfrac{1}{r}.$

Next, $pq + qr + rp = pqr \Rightarrow \dfrac{1}{r} + \dfrac{1}{p} + \dfrac{1}{q} = 1.$

So we get: $\dfrac{1}{p} + \dfrac{1}{q} + \dfrac{1}{r} = 1 \Rightarrow \log_t a + \log_t b + \log_t c = 1.$

Also, $\log_t a + \log_t b + \log_t c = \log_t abc$, too.

Therefore, $\log_t abc = 1 \Rightarrow abc = t^1 = t \Rightarrow t = abc.$

## Suggestions or Solutions
## To the Problem in the Example 3

Assuming $1 < c < d$, and $c^{\frac{1}{m}} = d^{\frac{1}{n}} = cd$, show that: $\sqrt{m^3 - m^2n - mn^2 + n^3} = \log_{dc}\frac{d}{c}$.

To begin with, $c^{\frac{1}{m}} = d^{\frac{1}{n}} = cd \Rightarrow \log c^{\frac{1}{m}} = \log d^{\frac{1}{n}} = \log cd \Rightarrow \frac{1}{m}\log c = \frac{1}{n}\log d = \log cd$.

So we get: $\frac{1}{m}\log c = \log cd$, and $\frac{1}{n}\log d = \log cd$.

We know that $cd$ is the base in the $\log_{dc}\frac{d}{c}$, so $cd \neq 1$, and thus, $\log cd \neq 0$.

Next, we get:

$\frac{1}{m}\log c = \log cd \Rightarrow \log c = m\log cd \Rightarrow m = \dfrac{\log c}{\log cd}$.

$\frac{1}{n}\log d = \log cd \Rightarrow \log d = n\log cd \Rightarrow n = \dfrac{\log d}{\log cd}$.

$m^3 - m^2n - mn^2 + n^3 = m^2(m - n) - n^2(m - n) = (m^2 - n^2)(m - n) = (m + n)(m - n)^2$.

We have: $m = \dfrac{\log c}{\log cd}$, and $n = \dfrac{\log d}{\log cd}$. Thus, we get:

$m + n = \dfrac{\log c}{\log cd} + \dfrac{\log d}{\log cd} = \dfrac{\log c + \log d}{\log cd} = \dfrac{\log cd}{\log cd} = 1$.

$m - n = \dfrac{\log c}{\log cd} - \dfrac{\log d}{\log cd} = \dfrac{\log c - \log d}{\log cd} = \dfrac{\log \frac{c}{d}}{\log cd} = \log_{cd}\frac{c}{d}$.

So we get: $m^3 - m^2n - mn^2 + n^3 = (m + n)(m - n)^2 = (m - n)^2$.

$1 < c < d \Rightarrow \log 1 < \log c < \log d \Rightarrow 0 < \log c < \log d \Rightarrow \log c - \log d < 0$.

$1 < c < d \Rightarrow 1 < cd$, since both $c$ and $d > 1$. So $1 < cd \Rightarrow \log 1 < \log cd \Rightarrow 0 < \log cd$.

Thus, we get: $m - n = \dfrac{\log c - \log d}{\log cd} < 0$.

So we get: $\sqrt{m^3 - m^2 n - mn^2 + n^3} = n - m = \dfrac{\log d - \log c}{\log cd} = \dfrac{\log \frac{d}{c}}{\log cd} = \log_{cd} \frac{d}{c}$.

Therefore, $\sqrt{m^3 - m^2 n - mn^2 + n^3} = \log_{dc} \frac{d}{c}$.

*If not quite sure of the idea behind the processes above, follow the steps below:*

Why do you do this example?

As many other examples in this book, this example, too, is for your practice on log algebra. That's simply because algebra matters. Following procedures and steps, or understanding processes when taking courses in math, particularly calculus, or other courses having much to do with math, you've got to do algebra.

Once understood ideas, formulas, identities, rules or laws in math, real caliber is up to the amount and quality of examples. Understanding is one thing, and doing it is another. Practice doesn't make things perfect, since nothing is perfect, but can make them better.

Now, what have we got in this example?

Finding $m$ and $n$, first, and then, putting them into the expression, $m^3 - m^2 n - mn^2 + n^3$, we should end up with the square of the right hand side, that is, we should get: $(\log_{dc} \frac{d}{c})^2$.

Finding $m$ and $n$, we want to solve $c^{\frac{1}{m}} = d^{\frac{1}{n}} = cd$, which is a system of two equations, one of which is: $c^{\frac{1}{m}} = cd$, and the other is: $d^{\frac{1}{n}} = cd$.

However, $m$ and $n$ are in the positions where exponents are.    What can we do then?

We can get them out of there taking logs. Usually, extracting an exponent from a power, we take a log of it (or apply a log to it). So applying logs to the expression, we get:

$$c^{\frac{1}{m}} = d^{\frac{1}{n}} = cd \Rightarrow \log c^{\frac{1}{m}} = \log d^{\frac{1}{n}} = \log cd \Rightarrow \tfrac{1}{m}\log c = \tfrac{1}{n}\log d = \log cd.$$

Thus, we get: $\frac{1}{m}\log c = \log cd$, and $\frac{1}{n}\log d = \log cd$.    So we get:

$$\frac{1}{m}\log c = \log cd \Rightarrow \frac{\log c}{m} = \log cd \Rightarrow \log c = m\log cd \Rightarrow m = \frac{\log c}{\log cd}.$$

And we get this, too:

$$\frac{1}{n}\log d = \log cd \Rightarrow \log d = n\log cd \Rightarrow n = \frac{\log d}{\log cd}.$$    What if $\log cd = 0$, though?

We know that $cd$ is the base in the $\log_{dc}\frac{d}{c}$, so $cd \neq 1$, and thus, $\log cd \neq 0$.

So $m$ and $n$ both are OK with the denominator $\log cd$.

Now, let's move on to the expression, $m^3 - m^2n - mn^2 + n^3$.

Before doing the substitutions however, we may want to examine the expression above.

That's because it seems that the substitutions will make the expression quite messy, so there probably be something we can do about the expression prior to the substitutions.

Closely looking at the expression, we can notice that it is quite symmetric, and can see that both $m$ and $n$ have the same situation. Therefore, we may want to try factorizing the expression. To begin with, we can put it the way below, too:

$$m^3 - m^2n - mn^2 + n^3 = m^3 - m^2n + n^3 - n^2m.$$

Factorizing the above, we get:

$$m^3 - m^2n + n^3 - n^2m = m^2(m-n) + n^2(n-m) = m^2(m-n) - n^2(m-n)$$
$$= (m^2 - n^2)(m-n) = (m+n)(m-n)(m-n) = (m+n)(m-n)^2,\text{ which looks quite}$$
simpler now.

So next, we may want to find $m + n$ and $m - n$, first.

We have: $m = \dfrac{\log c}{\log cd}$, and $n = \dfrac{\log d}{\log cd}$.   Thus, we get:

$$m + n = \frac{\log c}{\log cd} + \frac{\log d}{\log cd} = \frac{\log c + \log d}{\log cd} = \frac{\log cd}{\log cd} = 1.$$

$$m - n = \frac{\log c}{\log cd} - \frac{\log d}{\log cd} = \frac{\log c - \log d}{\log cd} = \frac{\log \frac{c}{d}}{\log cd} = \log_{cd} \frac{c}{d}.$$

So we get: $m^3 - m^2n - mn^2 + n^3 = (m + n)(m - n)^2 = (m - n)^2$, since $m + n = 1$.

Now, note that $\sqrt{m^3 - m^2n - mn^2 + n^3} \geq 0$, which is obvious, since a square root of any real number is greater than or equal to 0 if no minus sign is in front of it.

And what's inside the square root sign is: $\geq 0$, of course, since the root itself has to be real. That is, $\sqrt{m^3 - m^2n - mn^2 + n^3}$ is real if $m^3 - m^2n - mn^2 + n^3 \geq 0$.

Working with any power in this form: $b^{\frac{1}{n}}$ where $n$ is even, make sure that $b \geq 0$.

If $n$ is even and $b \geq 0$, we can set: $b^{\frac{1}{n}} = \sqrt[n]{b}$, where $\sqrt[n]{\phantom{x}}$ is called a radical sign, and is more specifically, called an $n^{\text{th}}$ root sign.

However, when we get to remove such a radical sign, we can make a mistake.

Removing such a radical sign when $n$ is even, we want to make sure what's inside the radical sign is $\geq 0$.

Suppose for instance, $y = \sqrt[4]{a^4}$.

Then, we can't just set: $y = a$ since we don't know if $a \geq 0$. So we want to set: $y = |a|$.

So suppose for instance, we have: $y = \sqrt[4]{a^4}$, where $a < 0$.

Then, we need to set: $y = -a$, because $y \geq 0$ and $a < 0$.

Now, we have: $\sqrt{m^3 - m^2n - mn^2 + n^3} = \sqrt{(m-n)^2} \geq 0$.

So we want to check to see if $m - n \geq 0$ before removing the radical sign (the square root sign) and the exponent, 2 from $(m-n)^2$.

If $m - n \geq 0$, we need to take $m - n$.

Otherwise, we want to take $n - m$ after the removal since $m - n < 0$.

That is:

If $m - n \geq 0$, we get: $\sqrt{(m-n)^2} = m - n$.

If $m - n < 0$, we get: $\sqrt{(m-n)^2} = -(m-n) = n - m$.

So let's give it a check now.    To begin with, we have: $1 < c < d$.

Thus, $1 < c < d \Rightarrow \log 1 < \log c < \log d \Rightarrow 0 < \log c < \log d \Rightarrow \log c - \log d < 0$.

On the other hand, we can get this, too: $1 < c < d \Rightarrow 1 < cd$, since both $c$ and $d > 1$.

So we get: $1 < cd \Rightarrow \log 1 < \log cd \Rightarrow 0 < \log cd$.

Thus, we can see that $m - n = \dfrac{\log c - \log d}{\log cd} < 0$, so we want to take $(n - m)$, which is $> 0$.

So we get: $n - m = \dfrac{\log d - \log c}{\log cd} = \dfrac{\log \frac{d}{c}}{\log cd} = \log_{cd} \frac{d}{c}$.

Thus, we get: $\sqrt{m^3 - m^2n - mn^2 + n^3} = n - m = \log_{cd} \frac{d}{c} = \log_{dc} \frac{d}{c}$, since $cd = dc$.

Therefore, $\sqrt{m^3 - m^2n - mn^2 + n^3} = \log_{dc} \frac{d}{c}$.

And we can put all the processes above in sum the way below:

To begin with, $c^{\frac{1}{m}} = d^{\frac{1}{n}} = cd \Rightarrow \log c^{\frac{1}{m}} = \log d^{\frac{1}{n}} = \log cd \Rightarrow \frac{1}{m}\log c = \frac{1}{n}\log d = \log cd$.

So we get: $\frac{1}{m}\log c = \log cd$, and $\frac{1}{n}\log d = \log cd$.

We know that $cd$ is the base in the $\log_{dc}\frac{d}{c}$, so $cd \neq 1$, and thus, $\log cd \neq 0$.

Next, we get:

$\frac{1}{m}\log c = \log cd \Rightarrow \log c = m\log cd \Rightarrow m = \dfrac{\log c}{\log cd}$.

$\frac{1}{n}\log d = \log cd \Rightarrow \log d = n\log cd \Rightarrow n = \dfrac{\log d}{\log cd}$.

$m^3 - m^2n - mn^2 + n^3 = m^2(m-n) - n^2(m-n) = (m^2 - n^2)(m-n) = (m+n)(m-n)^2$.

And we have: $m = \dfrac{\log c}{\log cd}$, and $n = \dfrac{\log d}{\log cd}$.    Thus, we get:

$m + n = \dfrac{\log c}{\log cd} + \dfrac{\log d}{\log cd} = \dfrac{\log c + \log d}{\log cd} = \dfrac{\log cd}{\log cd} = 1$.

$m - n = \dfrac{\log c}{\log cd} - \dfrac{\log d}{\log cd} = \dfrac{\log c - \log d}{\log cd} = \dfrac{\log \frac{c}{d}}{\log cd} = \log_{cd}\frac{c}{d}$.

So we get: $m^3 - m^2n - mn^2 + n^3 = (m+n)(m-n)^2 = (m-n)^2$.

$1 < c < d \Rightarrow \log 1 < \log c < \log d \Rightarrow 0 < \log c < \log d \Rightarrow \log c - \log d < 0$.

$1 < c < d \Rightarrow 1 < cd$, since both $c$ and $d > 1$. So $1 < cd \Rightarrow \log 1 < \log cd \Rightarrow 0 < \log cd$.

Thus, we get: $m - n = \dfrac{\log c - \log d}{\log cd} < 0$.

So we get: $\sqrt{m^3 - m^2n - mn^2 + n^3} = n - m = \dfrac{\log d - \log c}{\log cd} = \dfrac{\log \frac{d}{c}}{\log cd} = \log_{cd}\frac{d}{c}$.

Therefore, $\sqrt{m^3 - m^2n - mn^2 + n^3} = \log_{dc}\frac{d}{c}$.

## Examples 6 on Logarithms

In this set of examples, too, we are going to do some more practice on algebra on logs so that you get more familiar with manipulation of logs. That's because algebra matters.

0.  Assuming $0 < b < 1 < a$, show that $\log_a b < 0$.

1.  Assuming that $1 < b < a$, and $p = (\log_a b)^2$, $q = \log_a b^2$, and that $r = \log_a (\log_a b)$, put $p$, $q$, and $r$ in an ascending order. Note that $\log_a (\log_a b) = \log_a \log_a b$, so the use of parentheses in this case is merely for clarity purposes.

2.  Assuming $a^2 + b^2 = c^2$, show that $\log_{b+c} a + \log_{c-b} a = 2(\log_{b+c} a)(\log_{c-b} a)$.

## Suggestions or Solutions
## To the Problem in the Example 0

**Assuming $0 < b < 1 < a$, show that $\log_a b < 0$.**

Since $0 < b < 1 < a$, we get: $b < 1 \Rightarrow \log_a b < \log_a 1 = 0 \Rightarrow \log_a b < 0$.

We can put it this way, too:

$b < 1 \Rightarrow \log b < \log 1 = 0 \Rightarrow \log b < 0$. $1 < a \Rightarrow \log 1 < \log a \Rightarrow 0 < \log a$.

Therefore, $\dfrac{\log b}{\log a} < 0$, so we get: $\log_a b < 0$ since $\log_a b = \dfrac{\log b}{\log a}$.

*If not quite sure of the idea behind the processes above, follow the steps below:*

From the relational expression given, we can get two relational expressions as follows: $0 < b < 1$, and $1 < a$.

Using two expressions above, we should be able to draw the conclusion that $\log_a b < 0$. Of the two expressions, let's begin with the first.

The first one is: $0 < b < 1$, which means at the same time two cases as follows: $0 < b$ and $b < 1$.

Then first, since $b > 0$, we can take a log of $b$.

Next, let's move on to the expression, $b < 1$.

Suppose now, $1 < p$.

Then, applying a $\log_p$ to the expression $b < 1$, the direction of the inequality sign remains the same since the base $p > 1$. If $0 < p < 1$ however, the direction changes.

For instance, we have: $\frac{1}{2} > \frac{1}{8}$, and **0.04 < 0.008**.

Then, we get: $\log_2 \frac{1}{2} > \log_2 \frac{1}{8}$, because $\log_2 \frac{1}{2} = \log_2 2^{-1} = -1$, and $\log_2 \frac{1}{8} = \log_2 2^{-3} = -3$.

So the direction of the inequality sign remains the same if the base > 1.

And the same is true for any antilog positive, too.

For instance, $\log_2 8 > \log_2 4$, because $\log_2 8 = \log_2 2^3 = 3$, and $\log_2 4 = \log_2 2^2 = 2$.

However, we get: $\log_{0.2} 0.04 < \log_{0.2} 0.008$.

That's because $\log_{0.2} 0.04 = \log_{0.2} 0.2^2 = 2$, and $\log_{0.2} 0.008 = \log_{0.2} (0.2)^3 = 3$.

So the direction of the inequality sign changes if **0 < the base < 1**.

Now, let's move on to the second expression, $1 < a$.

Then, we get: $b < 1 \Rightarrow \log_a b < \log_a 1 \Rightarrow \log_a b < 0$.

Suppose however, that **0 < a < 1**, and **0 < b < 1**.

Then, we get: $b < 1 \Rightarrow \log_a b > \log_a 1 \Rightarrow \log_a b > 0$.

**In short:**

Since **0 < b < 1 < a**, we get: $b < 1 \Rightarrow \log_a b < \log_a 1 = 0 \Rightarrow \log_a b < 0$.

We can put it this way, too:

$b < 1 \Rightarrow \log b < \log 1 = 0 \Rightarrow \log b < 0.$ $1 < a \Rightarrow \log 1 < \log a \Rightarrow 0 < \log a$.

Therefore, $\dfrac{\log b}{\log a} < 0$, so we get: $\log_a b < 0$ since $\log_a b = \dfrac{\log b}{\log a}$.

## Suggestions or Solutions
## To the Problem in the Example 1

**Assuming that $1 < b < a$, and $p = (\log_a b)^2$, $q = \log_a b^2$, and that $r = \log_a (\log_a b)$, put $p$, $q$, and $r$ in an ascending order. Note that $\log_a (\log_a b) = \log_a \log_a b$, so the use of parentheses in this case is merely for clarity purposes.**

We have: $p = (\log_a b)^2$, $q = \log_a b^2 = 2\log_a b$, and $r = \log_a (\log_a b)$.

Besides, $1 < b < a \Rightarrow \log_a 1 < \log_a b < \log_a a \Rightarrow 0 < \log_a b < 1$.

Also, $1 < b < a \Rightarrow a > 1$.

So we now have: $a > 1$ and $0 < \log_a b < 1$.

Thus, $r = \log_a (\log_a b) < 0 \Rightarrow r < 0$.

Next, $q - p = 2 \log_a b - (\log_a b)^2 = (\log_a b)(2 - \log_a b)$.

Meanwhile, $0 < \log_a b < 1 \Rightarrow -1 < -\log_a b < 0 \Rightarrow 1 < 2 - \log_a b < 2$.

Thus, $(\log_a b)(2 - \log_a b) > 0 \Rightarrow q - p > 0 \Rightarrow q > p$.

Therefore, $q > p > r$.

*If not quite sure of the idea behind the processes above, follow the steps below:*

To begin with, let's take a closer look at the expressions given.
Then, we can notice that $p$, $q$, and $r$ are respectively put in terms of $\log_a b$.    How come?

In $\log_a (\log_a b)$, the antilog is $\log_a b$, and we can get: $\log_a b^2 = 2\log_a b$.

Thus, we may want to come up with something significant to $\log_a b$.   How?

We should be able to take advantage of the condition given.    What condition?

We are given a condition that $1 < b < a$, in which we should be able to find significant information on that log, which is $\log_a b$.

Then, out of the condition, we can come up with an expression with the log, $\log_a b$.

So let's first, take advantage of the condition, $1 < b < a$.

In the condition, we can see that $1 < a$.

Thus, if we apply a $\log_a$ to the condition, which is a relational expression, the directions of the inequality signs remain the same, so we get:

$1 < b < a \Rightarrow \log_a 1 < \log_a b < \log_a a \Rightarrow 0 < \log_a b < 1.$

Let's next, see what we can do about $p$, $q$ and $r$ respectively.

Then, setting $x = \log_a b$, we get: $p = x^2$, $q = 2x$, and $r = \log_a x$, in each of which $0 < x < 1$, since $x = \log_a b$, and $0 < \log_a b < 1$.

Let's look at $r$, first.

We know that the base $a > 1$.    So we get:

$\log_a x < 0$ if $0 < x < 1$.

$\log_a x \geq 0$ if $1 \leq x$.

Thus, we can see that $r = \log_a x < 0$, since the base $a > 1$ and $0 < x < 1$.   How come?

Refer to the problem in the Example 4.

Let's now, move on to $p = x^2$, and $q = 2x$, where $x = \log_a b$.

Then, we can readily see that both $p$ and $q > 0$ when $0 < x < 1$, so $r$ is the smallest of all since $r = \log_a x < 0$ when $0 < x < 1$.

Thus, we now, want to compare $p$ and $q$.

Comparing two objects, we can take either of the two ways as follows:
One is that we take the difference between the two, and then, check to see if it is positive.
The other is that we take the ratio between the two, and check to see if it is greater than 1.

And taking the difference, we get: $q - p = 2x - x^2 = x(2 - x) \Rightarrow q - p = x(2 - x)$.

We have: $0 < x < 1$, too.

Thus, we get: $0 < x < 1 \Rightarrow 0 > -x > -1 \Rightarrow 2 > (2 - x) > -1 + 2 = 1 \Rightarrow 1 < (2 - x) < 2$.

So we can see that both $x$ and $(2 - x)$ are positive at least.

Thus, $q - p = x(2 - x) > 0 \Rightarrow q - p > 0 \Rightarrow q > p$.

Now putting threads together, we get: $q > p > r$.

## Suggestions or Solutions
## To the Problem in the Example 2

**Suppose $a^2 + b^2 = c^2$. Then, show that $\log_{b+c} a + \log_{c-b} a = 2(\log_{b+c} a)(\log_{c-b} a)$.**

$$\log_{c+b} a + \log_{c-b} a = \frac{1}{\log_a c + b} + \frac{1}{\log_a c - b} = \frac{\log_a c - b + \log_a c + b}{(\log_a c + b)(\log_a c - b)} = \frac{\log_a (c - b)(c + b)}{(\log_a c + b)(\log_a c - b)}$$

$$= \frac{\log_a c^2 - b^2}{(\log_a c + b)(\log_a c - b)} = \frac{\log_a a^2}{(\log_a c + b)(\log_a c - b)} = \frac{2}{(\log_a c + b)(\log_a c - b)}. \text{ Meanwhile:}$$

$$\frac{1}{\log_a c + b} \cdot \frac{1}{\log_a c - b} = (\log_{c+b} a)(\log_{c-b} a). \text{ So } \log_{c+b} a + \log_{c-b} a = 2(\log_{c+b} a)(\log_{c-b} a).$$

*If not quite sure of the idea behind the processes above, follow the steps below:*

Closely looking at the equality in logs, we can see that all the logs have the same antilog, which is **a**, and are to bases same or similar. In fact, the equality is made of two different logs, which are $\log_{b+c} a$ and $\log_{c-b} a$. The only difference between the two is in the base.

Usually, it is a good idea to let logs have the same base. It is particularly the case if logs are in an expression as an equation, equality, etc. Now, all the logs in the equality above have the same antilog. So it is not going to be very hard to make the logs have the same base. The same base is going to be **a**, of course. We have a tool for such a work.

The tool is a log identity, where $\log_x y = \dfrac{1}{\log_y x}$ for $x$ and $y$ both $> 0$ but $\neq 1$.

So beginning with the left hand side in the equality, we get:

$$\log_{c+b} a + \log_{c-b} a = \frac{1}{\log_a c+b} + \frac{1}{\log_a c-b} = \frac{\log_a c-b+\log_a c+b}{(\log_a c+b)(\log_a c-b)} = \frac{\log_a (c-b)(c+b)}{(\log_a c+b)(\log_a c-b)}$$

$$= \frac{\log_a c^2 - b^2}{(\log_a c+b)(\log_a c-b)} = \frac{\log_a a^2}{(\log_a c+b)(\log_a c-b)} = \frac{2}{(\log_a c+b)(\log_a c-b)}.$$

Meanwhile, we have: $\dfrac{1}{\log_a c+b} \cdot \dfrac{1}{\log_a c-b} = (\log_{c+b} a)(\log_{c-b} a)$.

Thus, we get: $\log_{c+b} a + \log_{c-b} a = 2(\log_{c+b} a)(\log_{c-b} a)$, which is the equality we had to show.

In addition, considering the structure of the expression $a^2 + b^2 = c^2$ given as a condition in this problem, we can notice that the equality below can hold, too:

$\log_{c+a} b + \log_{c-a} b = 2(\log_{c+a} b)(\log_{c-a} b)$.

That's because even if $a$ and $b$ exchange their positions, the expression $a^2 + b^2 = c^2$ remains the same.   Not quite sure?

We have: $a^2 + b^2 = c^2 \Rightarrow \log_{c+b} a + \log_{c-b} a = 2(\log_{c+b} a)(\log_{c-b} a)$.

Now, if $a$ and $b$ exchange their positions, we get:

$a^2 + b^2 = c^2 \Rightarrow \log_{c+a} b + \log_{c-a} b = 2(\log_{c+a} b)(\log_{c-a} b)$.    Still not quite sure?

First, the equality $\log_{c+b} a + \log_{c-b} a = 2(\log_{c+b} a)(\log_{c-b} a)$ can hold if $a^2 + b^2 = c^2$.

Next, having $a$ and $b$ exchange their positions, we get:

$\log_{c+a} b + \log_{c-a} b = 2(\log_{c+a} b)(\log_{c-a} b)$ from $\log_{c+b} a + \log_{c-b} a = 2(\log_{c+b} a)(\log_{c-b} a)$, but still get $a^2 + b^2 = c^2$ from $a^2 + b^2 = c^2$.

That is, $a^2 + b^2 = c^2$ remains the same even if $a$ and $b$ exchange their positions.

So another equality $\log_{c+a} b + \log_{c-a} b = 2(\log_{c+a} b)(\log_{c-a} b)$ can hold, too, if $a^2 + b^2 = c^2$.

By the way, the equality $a^2 + b^2 = c^2$ can be called the right triangle identity, often called Pythagorean theorem, too, where the square of the hypotenuse equals the sum of the square of the base and that of the height. So in the equality $a^2 + b^2 = c^2$, $c$ can be taken for a hypotenuse, and $b$ and $c$ are line segments perpendicular to each other.

There is another way we can verify this: $\log_{b+c} a + \log_{c-b} a = 2(\log_{b+c} a)(\log_{c-b} a)$.

We put in a fraction every log in the equality above, and then, compare both sides of the equality. So beginning with the left hand side: $\log_{b+c} a + \log_{c-b} a$, and setting:

$$\log_{b+c} a = \frac{\log a}{\log(b+c)}, \text{ and } \log_{c-b} a = \frac{\log a}{\log(c-b)}, \text{ we get:}$$

$$\log_{b+c} a + \log_{c-b} a = \log a \left( \frac{1}{\log(c+b)} + \frac{1}{\log(c-b)} \right) = \log a \cdot \frac{\log(c-b) + \log(c+b)}{\log(c+b)\log(c-b)}.$$

So we get: $\log_{b+c} a + \log_{c-b} a = \log a \cdot \dfrac{\log(c-b) + \log(c+b)}{\log(c+b)\log(c-b)}.$

Meanwhile, we have:

$(\log a)\{\log (c - b) + \log (c + b)\} = (\log a)\{\log (c - b)(c + b)\} = (\log a)\{\log (c^2 - b^2)\}$

$= (\log a)(\log a^2) = (\log a)2(\log a) = 2(\log a)^2.$

Thus, we get: $\log_{b+c} a + \log_{c-b} a = \dfrac{2(\log a)^2}{\log(c+b)\log(c-b)}.$

Next, moving on to the right hand side, we get:

$$2(\log_{b+c} a)(\log_{c-b} a) = \frac{2\log a}{\log(b+c)} \cdot \frac{\log a}{\log(c-b)} = \frac{2(\log a)^2}{\log(c+b)\log(c-b)}, \text{ which is the same as}$$

the left hand side.

Let's this time, see if we can verify this: $\log_{c+a} b + \log_{c-a} b = 2(\log_{c+a} b)(\log_{c-a} b)$.

Beginning with the left hand side: $\log_{c+a} b + \log_{c-a} b$, and setting:

$$\log_{c+a} b = \frac{\log b}{\log(c+a)}, \text{ and } \log_{c-a} b = \frac{\log b}{\log(c-a)}, \text{we get:}$$

$$\log_{c+a} b + \log_{c-a} b = \log b \left( \frac{1}{\log(c+a)} + \frac{1}{\log(c-a)} \right) = \log b \cdot \frac{\log(c-a) + \log(c+a)}{\log(c+a)\log(c-a)}.$$

So we get: $\log_{c+a} b + \log_{c-a} b = \log b \cdot \dfrac{\log(c-a) + \log(c+a)}{\log(c+a)\log(c-a)}.$

Meanwhile, we have:

$(\log b)\{\log (c - a) + \log (c + a)\} = (\log b)\{\log (c - a)(c + a)\} = (\log b)\{\log (c^2 - a^2)\}$

$= (\log b)(\log b^2) = (\log b)2(\log b) = 2(\log b)^2.$

Thus, we get: $\log_{c+a} b + \log_{c-a} b = \dfrac{2(\log b)^2}{\log(c+a)\log(c-a)}.$

Next, moving on to the right hand side, we get:

$$2(\log_{c+a} b)(\log_{c-a} b) = \frac{2\log b}{\log(c+a)} \cdot \frac{\log b}{\log(c-a)} = \frac{2(\log b)^2}{\log(c+a)\log(c-a)}, \text{ which is the same as}$$

the left hand side.

**In short:**

$$\log_{c+b} a + \log_{c-b} a = \frac{1}{\log_a c+b} + \frac{1}{\log_a c-b} = \frac{\log_a c - b + \log_a c + b}{(\log_a c+b)(\log_a c-b)} = \frac{\log_a(c-b)(c+b)}{(\log_a c+b)(\log_a c-b)}$$

$$= \frac{\log_a c^2 - b^2}{(\log_a c+b)(\log_a c-b)} = \frac{\log_a a^2}{(\log_a c+b)(\log_a c-b)} = \frac{2}{(\log_a c+b)(\log_a c-b)}. \text{ Meanwhile:}$$

$$\frac{1}{\log_a c+b} \cdot \frac{1}{\log_a c-b} = (\log_{c+b} a)(\log_{c-b} a). \text{ So } \log_{c+b} a + \log_{c-b} a = 2(\log_{c+b} a)(\log_{c-b} a).$$

**The other method:**

On the left hand side of the equality, setting: $\log_{b+c} a = \frac{\log a}{\log(b+c)}$, and $\log_{c-b} a = \frac{\log a}{\log(c-b)}$,

we get: $\log_{b+c} a + \log_{c-b} a = \log a(\frac{1}{\log(c+b)} + \frac{1}{\log(c-b)}) = \log a \cdot \frac{\log(c-b)+\log(c+b)}{\log(c+b)\log(c-b)}$.

Meanwhile: $(\log a)\{\log (c - b) + \log (c + b)\} = (\log a)\{\log (c - b)(c + b)\}$

$= (\log a)\{\log (c^2 - b^2)\} = (\log a)(\log a^2) = (\log a)2(\log a) = 2(\log a)^2$.

Thus, we get: $\log_{b+c} a + \log_{c-b} a = \frac{2(\log a)^2}{\log(c+b)\log(c-b)}$.

Next, on the right hand side, we get:

$2(\log_{b+c} a)(\log_{c-b} a) = \frac{2\log a}{\log(b+c)} \cdot \frac{\log a}{\log(c-b)} = \frac{2(\log a)^2}{\log(c+b)\log(c-b)}$, which equals the left hand side.

Note that we have: $2(\log_{b+c} a)(\log_{b+c} a) = 2\log_{b+c} a \log_{b+c} a$, so the parentheses in this case are for clarity purposes only.

## Examples 7 on Logarithms

In this set of examples, too, we are going to do some more practice on algebra on logs so that you get more familiar with manipulation of logs. That's because algebra matters.

0.   Assuming $a$, $b$, $c$, and $k$ all are positive but not equal to 1, show that:
$2\log_b k = \log_a k + \log_c k \Leftrightarrow c^2 = (ac)^t$ where $t = \log_a b$.

1.   Solve simultaneous equations as follows: $\log_2 x + \log_3 y = 4$, & $(\log_3 x)(\log_2 y) = 3$.

## Suggestions or Solutions
## To the Problem in the Example 0

**Assuming $a$, $b$, $c$, and $k$ all are positive but not equal to 1, show that:**
$2\log_b k = \log_a k + \log_c k \Leftrightarrow c^2 = (ac)^t$ **where** $t = \log_a b$.

**Proof for:** $2\log_b k = \log_a k + \log_c k \Rightarrow c^2 = (ac)^t$ where $t = \log_a b$.

$$2\log_b k = \frac{2\log k}{\log b}.$$

$$\log_a k + \log_c k = \frac{\log k}{\log a} + \frac{\log k}{\log c} = \log k \frac{(\log c + \log a)}{(\log a)(\log c)} = \frac{(\log k)(\log ac)}{(\log a)(\log c)}.$$

Then, $2\log_b k = \log_a k + \log_c k \Rightarrow \dfrac{2\log k}{\log b} = \dfrac{(\log k)(\log ac)}{(\log a)(\log c)} \Rightarrow \dfrac{2}{\log b} = \dfrac{\log ac}{(\log a)(\log c)}.$

$$\Rightarrow 2\log c = \frac{(\log ac)(\log b)}{(\log a)} = \log ac \cdot \frac{\log b}{\log a} = (\log ac)(\log_a b).$$

$$\Rightarrow 2\log c = (\log ac)(\log_a b) \Rightarrow \log c^2 = (\log_a b)(\log ac).$$

Setting $t = \log_a b$, we get: $\log c^2 = t \log ac = \log (ac)^t$.

Therefore, $c^2 = (ac)^t$ where $t = \log_a b$.

**Proof for:** $c^2 = (ac)^t$ where $t = \log_a b \Rightarrow 2\log_b k = \log_a k + \log_c k$.

$$c^2 = (ac)^t \Rightarrow \log c^2 = \log (ac)^t = t \log ac = (\log_a b)(\log ac) = \frac{\log b}{\log a} \cdot \log ac$$

$$\Rightarrow \log c^2 = \frac{\log b}{\log a} \cdot \log ac \;\; \Rightarrow 2 \log c = \frac{\log b}{\log a} \cdot \log ac$$

$$\Rightarrow \frac{2}{\log b} = \frac{\log ac}{(\log a)(\log c)} \Rightarrow \frac{2 \log k}{\log b} = \frac{(\log k)(\log ac)}{(\log a)(\log c)}.$$

Then, we get: $2 \log_b k = \dfrac{(\log k)(\log c + \log a)}{(\log a)(\log c)} = \log k \left( \dfrac{\log c}{(\log a)(\log c)} + \dfrac{\log a}{(\log a)(\log c)} \right)$

$$= \log k \left( \frac{1}{\log a} + \frac{1}{\log c} \right) = \frac{\log k}{\log a} + \frac{\log k}{\log c} = \log_a k + \log_c k.$$

Therefore, $2\log_b k = \log_a k + \log_c k$.

So putting threads together, we have:

$2\log_b k = \log_a k + \log_c k \Rightarrow c^2 = (ac)^t$ where $t = \log_a b$.

$c^2 = (ac)^t$ where $t = \log_a b \Rightarrow 2\log_b k = \log_a k + \log_c k$.

Therefore, we get: $2\log_b k = \log_a k + \log_c k \Leftrightarrow c^2 = (ac)^t$ where $t = \log_a b$.

*If not quite sure of the idea behind the processes above, follow the steps below:*

This is just another example for practice on log algebra. Thus, doing this one, we get more used to such algebra. In this example, we can notice that if an equality is made of logs and all the logs have the same antilog, all the same antilogs can get canceled.

Besides, a log is an exponent, and thus, very naturally, can also, be an exponent in a power, of course. For instance, we can have $5^{\log 3}$.

Now, examining all expressions given in this problem, we can notice that the equality on the left hand side of $\Leftrightarrow$ has $k$s, but the one on the right hand side has none of those.

Thus, doing log algebra in this case, we will probably see all $k$s getting canceled.

Besides, we will notice that doing log algebra, we put a log in fractional form very often.

Let's now, begin with showing: $2\log_b k = \log_a k + \log_c k \Rightarrow c^2 = (ac)^t$ where $t = \log_a b$, and see how log algebra goes.

In other words, we break $(2\log_b k = \log_a k + \log_c k)$ into pieces, and then, put the pieces together to form $(c^2 = (ac)^t$ where $t = \log_a b)$.

To begin with, we have a log identity, $\log_x y = \dfrac{\log y}{\log x}$, where $x$ and $y$ both $> 0$ but $\neq 1$.
So we can get:

$$2\log_b k = 2 \cdot \frac{\log k}{\log b} = \frac{2\log k}{\log b}, \text{ and } \log_a k + \log_c k = \frac{\log k}{\log a} + \frac{\log k}{\log c} = \log k \left( \frac{1}{\log a} + \frac{1}{\log c} \right).$$

Meanwhile, $\dfrac{1}{\log a} + \dfrac{1}{\log c} = \dfrac{(\log c + \log a)}{(\log a)(\log c)} = \dfrac{\log ac}{(\log a)(\log c)}$.

Thus, we get: $\log_a k + \log_c k = \dfrac{(\log k)(\log ac)}{(\log a)(\log c)}$. And we have: $2\log_b k = \dfrac{2\log k}{\log b}$, too.

So we get: $2\log_b k = \log_a k + \log_c k \Rightarrow \dfrac{2\log k}{\log b} = \dfrac{(\log k)(\log ac)}{(\log a)(\log c)} \Rightarrow \dfrac{2}{\log b} = \dfrac{\log ac}{(\log a)(\log c)}$.

At this point, we may want to come up with $\log_a b$, which is in the expression $t = \log_a b$, which is the exponent on the right hand side in the equality $c^2 = (ac)^t$.

Now, we are showing that: $2\log_b k = \log_a k + \log_c k \Rightarrow c^2 = (ac)^t$ where $t = \log_a b$.

We now have: $2\log_b k = \log_a k + \log_c k \Rightarrow \dfrac{2}{\log b} = \dfrac{\log ac}{(\log a)(\log c)}$.

So what we want to show now is that: $\dfrac{2}{\log b} = \dfrac{\log ac}{(\log a)(\log c)} \Rightarrow c^2 = (ac)^t$.

In $c^2 = (ac)^t$, we have $c^2$ on the left hand side, and we have 2 in the numerator of $\dfrac{2}{\log b}$.

And we know: $2\log c = \log c^2$.

So multiplying by $(\log c)(\log b)$ each side in the equality $\dfrac{2}{\log b} = \dfrac{\log ac}{(\log a)(\log c)}$, we get:

$$2\log c = \frac{(\log ac)(\log b)}{(\log a)} = \log ac \cdot \frac{\log b}{\log a} = (\log ac)(\log_a b).$$

So we get: $2\log c = (\log ac)(\log_a b) \Rightarrow \log c^2 = (\log_a b)(\log ac)$.

And we know: $x \log_b A = \log_b A^x$.

So setting $t = \log_a b$, we get: $\log c^2 = t \log ac = \log (ac)^t \Rightarrow \log c^2 = \log (ac)^t$.

Therefore, we get: $c^2 = (ac)^t$ where $t = \log_a b$.

So we can now say that: $2\log_b k = \log_a k + \log_c k \Rightarrow c^2 = (ac)^t$ where $t = \log_a b$.

And next, we want to show: $c^2 = (ac)^t$ where $t = \log_a b \Rightarrow 2\log_b k = \log_a k + \log_c k$.

To begin with, taking logs of both sides in $c^2 = (ac)^t$, we get :

$$c^2 = (ac)^t \Rightarrow \log c^2 = \log (ac)^t = t \log ac = (\log_a b)(\log ac) = \frac{\log b}{\log a} \cdot \log ac$$

$$\Rightarrow \log c^2 = \frac{\log b}{\log a} \cdot \log ac \Rightarrow 2\log c = \frac{\log b}{\log a} \cdot \log ac.$$

Dividing both sides by $(\log b)(\log c)$ respectively, we get: $\dfrac{2}{\log b} = \dfrac{\log ac}{(\log a)(\log c)}$.

Multiplying both sides by $\log k$ respectively, we get: $\dfrac{2\log k}{\log b} = \dfrac{(\log k)(\log ac)}{(\log a)(\log c)}$.

And we know: $\dfrac{\log k}{\log b} = \log_b k$, and $\log ac = \log a + \log c$.

So we get: $2\log_b k = \dfrac{(\log k)(\log a + \log c)}{(\log a)(\log c)} = \log k\left(\dfrac{\log a}{(\log a)(\log c)} + \dfrac{\log c}{(\log a)(\log c)}\right)$

$= \log k\left(\dfrac{1}{\log c} + \dfrac{1}{\log a}\right) = \dfrac{\log k}{\log a} + \dfrac{\log k}{\log c} = \log_a k + \log_c k$.

Thus, we get: $2\log_b k = \log_a k + \log_c k$.

So we now have shown both below:

$2\log_b k = \log_a k + \log_c k \Rightarrow c^2 = (ac)^t$ where $t = \log_a b$.

$c^2 = (ac)^t$ where $t = \log_a b \Rightarrow 2\log_b k = \log_a k + \log_c k$.

Therefore, we get: $2\log_b k = \log_a k + \log_c k \Leftrightarrow c^2 = (ac)^t$ where $t = \log_a b$.

## Suggestions or Solutions
## To the Problem in the Example 1

**Solve simultaneous equations as follows: $\log_2 x + \log_3 y = 4$, and $(\log_3 x)(\log_2 y) = 3$.**

$(\log_3 x)(\log_2 y) = 3 \Rightarrow (\log_2 x)(\log_3 y) = 3$.

Thus, we get: $\log_2 x + \log_3 y = 4$, and $(\log_2 x)(\log_3 y) = 3$.

Setting: $u = \log_2 x$, and $v = \log_3 y$, we get a system where $u + v = 4$ and $uv = 3$.

So $u$ and $v$ can be taken for the roots of an equation $x^2 - 4x + 3 = 0$.

$x^2 - 4x + 3 = (x - 3)(x - 1) = 0 \Rightarrow x = 3$ or $1$.

Thus, we get: $(u, v) = (3, 1)$ or $(1, 3)$.

So we get:

$\log_2 x = 3 \Rightarrow x = 2^3 = 8$, and $\log_3 y = 1 \Rightarrow y = 3$.

$\log_2 x = 1 \Rightarrow x = 2$, and $\log_3 y = 3 \Rightarrow y = 3^3 = 27$.

Therefore, $(x, y) = (8, 3)$ or $(2, 27)$.

*If not quite sure of the idea behind the processes above, follow the steps below:*

All the tools in math are efficient. Of all the tools in math, substitutions are probably the most efficient.

In the system of equations given above, each unknown is an antilog in a log.

That is, unknowns are inside logs.   How then, do we get access to the unknowns?

Getting the access, we don't need to isolate the unknowns. In fact, we may not want to try getting direct access to those. We can get indirect access, then get the values of the unknowns.   That is to say that we may want to consult substitutions.   How though?

To begin with, we take for an unknown, $\log_2 x$ rather than $x$ itself.
Next, we do the same to the others, too.

Then, we solve the system, and then, get the values of $x$ and $y$.
So for instance, we can set: $u = \log_2 x$, and then, move on to the algebra.

That's just one thing, though, in this case.
Another is that we have four different kinds. That is, we have for different logs.

Doing such substitutions, we get to have not two unknowns only, but actually have four, which are $\log_2 x$, $\log_3 y$, $\log_3 x$, and $\log_2 y$.

Well then, we force them to be two only.   How though?

We can use another tool, which is as follows: $(\log_s P)(\log_t Q) = (\log_t P)(\log_s Q)$.

So we can swap the bases when two logs are multiplied by each other.

Let's first have a look at though, how we can get such a tool.

We have a log identity where $\log_x y = \dfrac{\log_c y}{\log_c x}$.

Thus, we get: $\log_s P = \dfrac{\log_t P}{\log_t s}$, and $\log_t Q = \dfrac{\log_s Q}{\log_s t}$.

Meanwhile, $\log_s t = \dfrac{1}{\log_t s} \Rightarrow \log_t s = \dfrac{1}{\log_s t}$.

Thus, we get: $\log_t Q = \dfrac{\log_s Q}{\log_s t} = (\log_s Q)(\log_t s)$.  And we have: $\log_s P = \dfrac{\log_t P}{\log_t s}$, too.

So we get: $(\log_s P)(\log_t Q) = \dfrac{\log_t P}{\log_t s}(\log_s Q)(\log_t s) = (\log_t P)(\log_s Q)$.

Therefore, we get: $(\log_3 x)(\log_2 y) = 3 \Rightarrow (\log_2 x)(\log_3 y) = 3$.

That is to say that we can swap the bases, which we can do of course, when two logs are multiplied by each other.

Now, we have: $\log_2 x + \log_3 y = 4$, and $(\log_2 x)(\log_3 y) = 3$.

So setting: $u = \log_2 x$, and $v = \log_3 y$, we get a system of $u + v = 4$ and $uv = 3$.

Solving for $u$ the first of the two equations, we get: $u = 4 - v$.

Next, putting $4 - v$ into $u$ in $uv = 3$, we get an equation for $v$ only, which is quadratic.

Then, we get to solve for $v$ the quadratic equation.

And we can put the same the way below, too, of course:

We can take $u$ and $v$ for the solutions to a quadratic equation: $(x - u)(x - v) = 0$.

Then, expanding the equation above, we get: $(x - u)(x - v) = x^2 - (u + v)x + uv = 0$.

And we know: $u + v = 4$ and $uv = 3$.

So $u$ and $v$ are the roots of a quadratic equation, $x^2 - 4x + 3 = 0$.

Therefore, we should get the same equation via the following procedure, too.

$u = 4 - v \Rightarrow uv = 3 \Rightarrow (4 - v)v = 3 \Rightarrow 4v - v^2 = 3 \Rightarrow v^2 - 4v + 3 = 0.$

What then, about $u$?

The same is true for $u$, too.

Since $u$ and $v$ are in the same situation in the system, we will get: $u^2 - 4u + 3 = 0$.

Thus, one of the two roots is the value of $u$, and the other is the value of $v$.

What do we mean by the same situation, though?

Even if $u$ and $v$ exchange their positions in the system where $u + v = 4$ and $uv = 3$, the system remains the same. So let's now, get the solution to $x^2 - 4x + 3 = 0$. Then, we get:

$x^2 - 4x + 3 = (x - 3)(x - 1) = 0 \Rightarrow x = 3$ or $1$.

Now, we know $x$ is actually acting for $u$ or $v$.

Thus, we have two cases as follows: $u = 3$ and $v = 1$, or $u = 1$ and $v = 3$.

In sum, we can set: $(u, v) = (3, 1)$ or $(1, 3)$.

We have: $u = \log_2 x$, and $v = \log_3 y$.

Thus, we get:

$\log_2 x = 3 \Rightarrow x = 2^3 = 8$, and $\log_3 y = 1 \Rightarrow y = 3^1 = 3$.

$\log_2 x = 1 \Rightarrow x = 2^1 = 2$, and $\log_3 y = 3 \Rightarrow y = 3^3 = 27$.

Therefore, $(x, y) = (8, 3)$ or $(2, 27)$.

## Examples 8 on Logarithms

0.   Assuming $n$ is integer, and $1.25^n$ has 8 digits above the decimal point, find $n$. Use $\log 2 = 0.3010$.

1.   Suppose that $a$ and $b$ are positive integers, and that $a^{100}$ has 120 digits. Then:

1.0.   Find the highest place value in $a^{-1}$.

1.1.   Find the value of $b$ that makes $a^b$ a 10-digit integer.

## Suggestions or Solutions
## To the Problem in the Example 0

**Assuming $n$ is integer, and $1.25^n$ has 8 digits above the decimal point, find $n$. Use $\log 2 = 0.3010$.**

$\log 1.25^n = n \log 1.25 = 7 + m$, where $0 \leq m < 1$.

$\log 1.25 = \log \frac{125}{100} = \log \frac{1000}{800} = \log \frac{10}{8} = 1 - 3 \log 2 = 1 - 3 \cdot 0.3010 = 1 - 0.9030 = 0.097$.

Thus, $n \log 1.25 = 0.097n = 7 + m$.

Besides, we have $0 \leq m < 1$, too.

Thus, $7 \leq 7 + m < 8 \Rightarrow 7 \leq 0.097n < 8 \Rightarrow \frac{7}{0.097} \leq n < \frac{8}{0.097} \Rightarrow 72 + \frac{16}{97} \leq n < 82 + \frac{46}{97}$.

Therefore, $n = 73, 74, 75, ...,$ and $82$.

*If not quite sure of the idea behind the processes above, follow the steps below:*

What does the number of digits in a number have to do with?

It has much to do with the highest place value in the number. How then, can we get the place value?

We can get the information on it by taking a common log of the number.

Assuming the number is $A$, and taking a common log of it, we can set:

$\log A = c + m$, where $c$ is an integer, and $0 \leq m < 1$.

Then, $c$ is often called the characteristic, $m$ is called the mantissa, and the highest place value in $A$ is $10^c$.

So if we have: $A \geq 1$, we get: $c \geq 0$, and the number of digits above the decimal point in $A$ is: $c + 1$.

If however, we have: $0 < A < 1$, we get: $c < 0$, so the first nonzero appears at the $|c|^{th}$ digit below the decimal point in $A$. For instance, if $c = -3$, the first nonzero appears at the $3^{rd}$ digit below the decimal point.

Let's now, begin with taking a common log of the number given, which is $1.25^n$.

Since the number, $1.25^n$ has 8 digits above the point, the characteristic $c = 7$.

Thus, we can set: $\log 1.25^n = n \log 1.25 = 7 + m$, where $0 \leq m < 1$.

So next, we want to get the value of $\log 1.25$. So can we get the value from a log table?

No, we are not allowed to use a log table doing this problem. What else then, can we do?

We are given the value of $\log 2$, which is 0.3010, so we should be able to put $\log 1.25$ in terms of $\log 2$. So we've got to do some log algebra, and doing it, we can get:

$1.25 = \frac{125}{100} = \frac{5}{4} = \frac{10}{8} \Rightarrow \log 1.25 = \log \frac{10}{8} = 1 - 3 \log 2 = 1 - 3 \cdot 0.3010 = 1 - 0.9030 = 0.097.$

Thus, we get: $n \log 1.25 = 7 + m \Rightarrow 0.097n = 7 + m$.

Besides, we have: $0 \leq m < 1$, too. So we get: $7 \leq 7 + m < 8$.

Thus, we get: $7 \leq 0.097n < 8 \Rightarrow \frac{7}{0.097} \leq n < \frac{8}{0.097} \Rightarrow 72 + \frac{16}{97} \leq n < 82 + \frac{46}{97}$.

Therefore, $n = 73, 74, 75, ...,$ and $82$.

Of course, we can begin with setting the relational expression as follows:

$10^7 \leq 1.25^n < 10^8$.   How come?

Since the number of digits above the decimal point in $1.25^n$ is 8, the highest place value in it is $10^7$.

Thus, we can set: $10^7 \leq 1.25^n < 10^8$.

Then, we can take the common log of each term in the expression above, and then, do the log algebra. So taking the logs, then doing the algebra, we get:

$10^7 \leq 1.25^n < 10^8 \Rightarrow \log 10^7 \leq \log 1.25^n < \log 10^8 \Rightarrow 7 \leq \log 1.25^n < 8$
$\Rightarrow 7 \leq n \log 1.25 < 8$.

Next, gettig the value of $\log 1.25$, we've got to do some more log algebra, and doing it, we can get:

$\log 1.25 = \log \frac{125}{100} = \log \frac{1000}{800} = \log \frac{10}{8} = 1 - 3 \log 2 = 1 - 3 \cdot 0.3010 = 1 - 0.9030 = 0.097$.

Thus, we get: $7 \leq 0.097n < 8 \Rightarrow \frac{7}{0.097} \leq n < \frac{8}{0.097} \Rightarrow 72 + \frac{16}{97} \leq n < 82 + \frac{46}{97}$.

Therefore, $n = 73, 74, 75, \ldots$, and $82$.

**In short:**

$\log 1.25^n = n \log 1.25 = 7 + m$, where $0 \leq m < 1$.

$\log 1.25 = \log \frac{125}{100} = \log \frac{1000}{800} = \log \frac{10}{8} = 1 - 3 \log 2 = 1 - 3 \cdot 0.3010 = 1 - 0.9030 = 0.097$.

Thus, $n \log 1.25 = 0.097n = 7 + m$.   Besides, we have $0 \leq m < 1$, too.

Thus, $7 \leq 7 + m < 8 \Rightarrow 7 \leq 0.097n < 8 \Rightarrow \frac{7}{0.097} \leq n < \frac{8}{0.097} \Rightarrow 72 + \frac{16}{97} \leq n < 82 + \frac{46}{97}$.

Therefore, $n = 73, 74, 75, \ldots$, and $82$.

## Suggestions or Solutions
## To the Problem 0 in the Example 1

**Assuming that $a$ and $b$ are positive integers, and that $a^{100}$ has 120 digits, find the highest place value in a number, $a^{-1}$.**

$\log a = 119 + m$, and $0 \le m < 1$.

$0 \le m < 1 \Rightarrow 119 \le 119 + m < 120 \Rightarrow 119 \le 100 \log a < 120 \Rightarrow 1.19 \le \log a < 1.2$

$\Rightarrow -1.2 < -\log a \le -1.19 \Rightarrow -1.2 < \log a^{-1} \le -1.19 \Rightarrow -1 - 0.2 < \log a^{-1} \le -1 - 0.19$

$\Rightarrow -2 + 0.8 < \log a^{-1} \le -2 + 0.81.$

So the characteristic is -2, and therefore, the highest place value is $10^{-2}$.

*If not quite sure of the idea behind the processes above, follow the steps below:*

Since $a^{100}$ has 120 digits, we can set: $\log a^{100} = 119 + m$, where $0 \le m < 1$.

Thus, we get: $100 \log a = 119 + m$, where $0 \le m < 1$.

Now, we've got to do some log algebra. What then, do we begin the algebra with?

We have: $0 \le m < 1$, so let's begin with it.

$0 \le m < 1 \Rightarrow 119 \le 119 + m < 120 \Rightarrow 119 \le 100 \log a < 120 \Rightarrow 1.19 \le \log a < 1.2$
$\Rightarrow -1.2 < -\log a \le -1.19 \Rightarrow -1.2 < \log a^{-1} \le -1.19 \Rightarrow -1 - 0.2 < \log a^{-1} \le -1 - 0.19.$

Why do we do it that way, though?

That's because the mantissa is positive, and we want to get the characteristic, too.

Thus, getting the characteristic, too, we get:

$-1 - 1 + 1 - 0.2 < \log a^{-1} \leq -1 - 1 + 1 - 0.19 \Rightarrow -2 + 0.8 < \log a^{-1} \leq -2 + 0.81$.

So the characteristic is -2.

Therefore, the highest place value in $a^{-1}$ is $10^{-2}$.

In other words, it has one zero before the first nonzero digit appears below the decimal point.

That is, the first nonzero digit appears in the second digit below the point.

**In short:**

$\log a = 119 + m$, and $0 \leq m < 1$.

$0 \leq m < 1 \Rightarrow 119 \leq 119 + m < 120 \Rightarrow 119 \leq 100 \log a < 120 \Rightarrow 1.19 \leq \log a < 1.2$

$\Rightarrow -1.2 < -\log a \leq -1.19 \Rightarrow -1.2 < \log a^{-1} \leq -1.19 \Rightarrow -1 - 0.2 < \log a^{-1} \leq -1 - 0.19$

$\Rightarrow -2 + 0.8 < \log a^{-1} \leq -2 + 0.81$.

So the characteristic is -2, and therefore, the highest place value is $10^{-2}$.

## Suggestions or Solutions
## To the Problem 1 in the Example 1

**Assuming that $a$ and $b$ are positive integers, and that $a^{100}$ has 120 digits, find the value of $b$ that makes $a^b$ a 10-digit integer.**

$\log a^b = b \log a = 9 + m$, where $0 \le m < 1$.

$0 \le m < 1 \Rightarrow 9 \le 9 + m < 10 \Rightarrow 9 \le b \log a < 10$.

$100 \log a = 119 + m$ where, $0 \le m < 1$.

$0 \le m < 1 \Rightarrow 119 \le 119 + m < 120 \Rightarrow 119 \le 100 \log a < 120 \Rightarrow 1.19 \le \log a < 1.2$
$\Rightarrow 1.19b \le b \log a < 1.2b$.

Besides, $9 \le b \log a < 10$, too. So $9 \le 1.19b \le b \log a < 1.2b \le 10$.

Then, first, we get: $9 \le 1.19b \Rightarrow \frac{9}{1.19} \le b \Rightarrow \frac{9}{1.19} = \frac{900}{119} = 7 + \frac{67}{119} \le b$.

Next, we get: $1.2b \le 10 \Rightarrow b \le \frac{10}{1.2} = \frac{100}{12} = 8 + \frac{4}{12} = 8 + \frac{1}{3}$.

So we get: $7 + \frac{67}{119} \le b \le 8 + \frac{1}{3} \Rightarrow b = 8$, since $b$ is an integer.

*If not quite sure of the idea behind the processes above, follow the steps below:*

Since $a^b$ is a 10-digit integer, the highest place value in it is $10^9$.

Thus, we can set: $\log a^b = 9 + m$, where $0 \le m < 1$, and we get: $b \log a = 9 + m$.

So we can get the extent of $b \log a$. How?

We have: $0 \le m < 1$, so we can get: $0 \le m < 1 \Rightarrow 9 \le 9 + m < 10 \Rightarrow 9 \le b \log a < 10$.

Now, we need the value of **log $a$**.    How then, can we get it?

Since $a^{100}$ has 120 digits, we can set: **100 log $a$ = 119 + $m$**, where **$0 \leq m < 1$**. So:

$$0 \leq m < 1 \Rightarrow 119 \leq 119 + m < 120 \Rightarrow 119 \leq 100 \log a < 120 \Rightarrow 1.19 \leq \log a < 1.2$$

Thus, we get: **$1.19b \leq b \log a < 1.2b$**.

Now, we have: **$9 \leq b \log a < 10$**, and **$1.19b \leq b \log a < 1.2b$**.

So we need to have: **$9 \leq 1.19b$**, and **$1.2b \leq 10$**.

In sum, we have: **$9 \leq 1.19b \leq b \log a < 1.2b \leq 10$**.

Thus, first, we get: $9 \leq 1.19b \Rightarrow \dfrac{9}{1.19} \leq b \Rightarrow \dfrac{9}{1.19} = \dfrac{900}{119} = 7 + \dfrac{67}{119} \leq b$.

Next, we get: $1.2b \leq 10 \Rightarrow b \leq \dfrac{10}{1.2} = \dfrac{100}{12} = 8 + \dfrac{4}{12} \Rightarrow b \leq 8 + \dfrac{1}{3}$.

Thus, we get: $7 + \dfrac{67}{119} \leq b \leq 8 + \dfrac{1}{3} \Rightarrow b = 8$, since $b$ is an integer.

## Examples 9 on Logarithms

0.   Assuming $\dfrac{x(y+z-x)}{\log_a x} = \dfrac{y(z+x-y)}{\log_a y} = \dfrac{z(x+y-z)}{\log_a z}$, show that $y^z z^y = z^x x^z = x^y y^x$.

1.   Assuming $\log_{2a} a = x$, and $\log_{3a} 2a = y$, show that $2^{1-xy} = 3^{y-xy}$.

2.   Assuming $3^x = 4^y = 6^z$, show that $\dfrac{1}{2y} = \dfrac{1}{z} - \dfrac{1}{x}$.

## Suggestions or Solutions
## To the Problem in the Example 0

**Assuming** $\dfrac{x(y+z-x)}{\log_a x} = \dfrac{y(z+x-y)}{\log_a y} = \dfrac{z(x+y-z)}{\log_a z}$, **show that** $y^z z^y = z^x x^z = x^y y^x$.

Setting: $\dfrac{x(y+z-x)}{\log_a x} = \dfrac{y(z+x-y)}{\log_a y} = \dfrac{z(x+y-z)}{\log_a z} = \dfrac{1}{m}$, we get:

$\log_a x = mx(y+z-x)$, $\log_a y = my(z+x-y)$, and $\log_a z = mz(x+y-z)$.

So we get: $\log_a x = mx(y+z-x) \Rightarrow x = a^{mx(y+z-x)}$. Thus, we get:

$x^z = (a^{mx(y+z-x)})^z = a^{mxz(y+z-x)}$, and $x^y = (a^{mx(y+z-x)})^y = a^{mxy(y+z-x)}$.

We know the same is true for $y$ and $z$, too, since they are in the same situation as $x$ is in. Thus, we get:

$y^z = (a^{my(z+x-y)})^z = a^{myz(z+x-y)}$, and $y^x = (a^{my(z+x-y)})^x = a^{myx(z+x-y)}$.

$z^x = (a^{mz(x+y-z)})^x = a^{mzx(x+y-z)}$, and $z^y = (a^{mz(x+y-z)})^y = a^{mzy(x+y-z)}$.

So we get: $x^y y^x = a^{mxy(y+z-x)} a^{myx(z+x-y)} = a^{mxy(y+z-x)+myx(z+x-y)}$.

Looking at the exponent part only, we can see that:

$mxy(y+z-x) + myx(z+x-y) = mxyy + mxyz - mxyx + mxyz + myxx - myxy = 2mxyz.$

Thus, we get: $x^y y^x = a^{2mxyz}$.

Since all the variables are in the same situation, the same is true for $y^z z^y$ and $z^x x^z$, too.

Thus, we get: $y^z z^y = a^{2mxyz}$, and $z^x x^z = a^{2mxyz}$.

Therefore, $x^y y^x = y^z z^y = z^x x^z$.

*If not quite sure of the idea behind the processes above, follow the steps below:*

This is just another example on log algebra, too. So we have nothing to memorize here, and we will just get more used to manipulations of algebra with logs.

Examining the equality we are to prove, we can see that it is in terms of $x$, $y$, and $z$ only with no log at all. Thus, we may want to extract $x$, $y$, and $z$ from the assumption.

Besides, we can notice that all the variables are in the same situation.

Thus, we may want to concentrate one of them, first, since the same idea will probably apply to the others.    Where do we begin, though?

We can put the problem this way, too:

$$\frac{x(y+z-x)}{\log_a x} = \frac{y(z+x-y)}{\log_a y} = \frac{z(x+y-z)}{\log_a z} \implies y^z z^y = z^x x^z = x^y y^x.$$

So we may want to begin with the assumption given.

What have we got in the assumption, though?

We've got just a bunch of complicated expressions connected by equal signs, don't we?

So it looks quite messy. It's not too bad, though. It seems quite involved, but eventually, will reduce to a ratio. In fact, it is nothing but a ratio. So we've got a ratio to work with.

We can simplify it to be in a form of $\frac{A}{B} = \frac{C}{D} = \frac{E}{F}$.    Why such a form, though?

It says all the three ratios are the same.

It doesn't specify what they are same to be, though.    Then, what do we do?

Such a ratio as above can be taken for a number as 3, 0.5, 8, 1.2, -1, etc.

For instance, we can have: $\frac{2}{4} = \frac{1}{2} = 0.5$, $\frac{-6}{4} = \frac{-3}{2} = -1.5$, $\frac{12}{4} = \frac{36}{12} = 3$, etc.

So we can set to a number the ratio above so that we can refer to the ratio more readily.

We don't know what the number is, though. Let's then, set it to a constant $m$.

So suppose now, $\dfrac{x(y+z-x)}{\log_a x} = \dfrac{y(z+x-y)}{\log_a y} = \dfrac{z(x+y-z)}{\log_a z} = m$.

Then, we may want to expand a little bit the ratio above to see if such a setting will let us readily extract $x$, $y$, and $z$. So expanding the first part above, we get:

$$\frac{x(y+z-x)}{\log_a x} = m \implies m \log_a x = x(y+z-x) \implies \log_a x = \frac{x}{m}(y+z-x).$$

Then, $m$ is in the denominator, and thus, can cause further calculations to be messier.

So we may want to set the ratio differently, and for instance, we can try $\frac{1}{m}$, where $m \neq 0$, of course. What if it doesn't work, either?

Then, we want to try something else, of course.

Now, setting $\dfrac{x(y+z-x)}{\log_a x} = \dfrac{y(z+x-y)}{\log_a y} = \dfrac{z(x+y-z)}{\log_a z} = \dfrac{1}{m}$, we get:

$\log_a x = mx(y+z-x)$, $\log_a y = my(z+x-y)$, and $\log_a z = mz(x+y-z)$.

Let's focus on $x$ only, for now.

Then, we get: $\log_a x = mx(y+z-x) \implies x = a^{mx(y+z-x)}$ by the definition for logs.

Now, we have $x^z$ and $x^y$ in the equality where $y^z z^y = z^x x^z = x^y y^x$.

So we may want to make it that way.   Then, we get:

$$x = a^{mx(y+z-x)} \Rightarrow x^z = (a^{mx(y+z-x)})^z = a^{mxz(y+z-x)}.$$

$$x = a^{mx(y+z-x)} \Rightarrow x^y = (a^{mx(y+z-x)})^y = a^{mxy(y+z-x)}.$$

We know the same is true for $y$ and $z$, too, since they are in the same situation as $x$ is in. Thus, we get:

$$y^z = (a^{my(z+x-y)})^z = a^{myz(z+x-y)}.$$

$$y^x = (a^{my(z+x-y)})^x = a^{myx(z+x-y)}.$$

$$z^x = (a^{mz(x+y-z)})^x = a^{mzx(x+y-z)}.$$

$$z^y = (a^{mz(x+y-z)})^y = a^{mzy(x+y-z)}.$$

Now, all we have to do is to make each of $y^z z^y$, $z^x x^z$, and $x^y y^x$, and show that they are all equal to each other.

To begin with, we get: $x^y y^x = a^{mxy(y+z-x)} a^{myx(z+x-y)} = a^{mxy(y+z-x) + myx(z+x-y)} = a^{2mxyz}.$ How come?

Looking at the exponent part only, we can see that:

$$mxy(y+z-x) + myx(z+x-y) = mxyy + mxyz - mxyx + mxyz + myxx - myxy = 2mxyz.$$

Since all the variables are in the same situation, the same is true for $y^z z^y$ and $z^x x^z$, too.

Thus, we get: $y^z z^y = a^{2mxyz}$, and $z^x x^z = a^{2mxyz}.$

Therefore, $x^y y^x = y^z z^y = z^x x^z.$

**In short:**

Setting: $\dfrac{x(y+z-x)}{\log_a x} = \dfrac{y(z+x-y)}{\log_a y} = \dfrac{z(x+y-z)}{\log_a z} = \dfrac{1}{m}$, we get:

$\log_a x = mx(y+z-x)$, $\log_a y = my(z+x-y)$, and $\log_a z = mz(x+y-z)$.

So we get: $\log_a x = mx(y+z-x) \Rightarrow x = a^{mx(y+z-x)}$.   Thus, we get:

$x^z = (a^{mx(y+z-x)})^z = a^{mxz(y+z-x)}$, and $x^y = (a^{mx(y+z-x)})^y = a^{mxy(y+z-x)}$.

We know the same is true for $y$ and $z$, too, since they are in the same situation as $x$ is in. Thus, we get:

$y^z = (a^{my(z+x-y)})^z = a^{myz(z+x-y)}$, and $y^x = (a^{my(z+x-y)})^x = a^{myx(z+x-y)}$.

$z^x = (a^{mz(x+y-z)})^x = a^{mzx(x+y-z)}$, and $z^y = (a^{mz(x+y-z)})^y = a^{mzy(x+y-z)}$.

So we get: $x^y y^x = a^{mxy(y+z-x)} a^{myx(z+x-y)} = a^{mxy(y+z-x)+myx(z+x-y)}$.

Looking at the exponent part only, we can see that:

$mxy(y+z-x) + myx(z+x-y) = mxyy + mxyz - mxyx + mxyz + myxx - myxy = 2mxyz$.

Thus, we get: $x^y y^x = a^{2mxyz}$.

Since all the variables are in the same situation, the same is true for $y^z z^y$ and $z^x x^z$, too.

Thus, we get: $y^z z^y = a^{2mxyz}$, and $z^x x^z = a^{2mxyz}$.

Therefore, $x^y y^x = y^z z^y = z^x x^z$.

## Suggestions or Solutions
## To the Problem in the Example 1

**Assuming $\log_{2a} a = x$, and $\log_{3a} 2a = y$, show that $2^{1-xy} = 3^{y-xy}$.**

$$x = \log_{2a} a = \frac{\log a}{\log 2a}, \text{ and } y = \log_{3a} 2a = \frac{\log 2a}{\log 3a}.$$

Thus, $xy = \dfrac{\log a}{\log 3a} = \log_{3a} a.$

$1 - xy = \log_{3a} 3a - \log_{3a} a = \log_{3a} \frac{3a}{a} = \log_{3a} 3 \Rightarrow 3 = (3a)^{1-xy}.$

$y - xy = \log_{3a} 2a - \log_{3a} a = \log_{3a} \frac{2a}{a} = \log_{3a} 2 \Rightarrow 2 = (3a)^{y-xy}.$

$3 = (3a)^{1-xy} \Rightarrow 3^{y-xy} = \{(3a)^{1-xy}\}^{y-xy}.$

$2 = (3a)^{y-xy} \Rightarrow 2^{1-xy} = \{(3a)^{y-xy}\}^{1-xy} = \{(3a)^{1-xy}\}^{y-xy}.$

Therefore, $2^{1-xy} = 3^{y-xy}.$

*If not quite sure of the idea behind the processes above, follow the steps below:*

This is just another example on log algebra, too.

Looking at the equality we are to prove, we see no log, but can see that both sides are antilogs, where the bases are two numbers, 2 and 3, but the exponents are expressions in terms of *x* and *y*.

Examining the assumption though, we can see *x* and *y* are all expressed in terms of logs.

In such cases, we usually remove the log signs, extract the bases necessary, which are 2 and 3, and then, form the equality, which is to be proved, of course.

Anyhow, log signs will probably disappear during such a forming process.

So let's now, begin extracting the two bases, 2 and 3.

We know a log can be put in a fractional form.
Quite often, we in fact, take a log for a fraction, where a numerator is over a denominator.

Thus, we may want to begin with putting the two logs in fractional forms. Then, we get:

$$x = \log_{2a} a = \frac{\log a}{\log 2a}, \text{ and } y = \log_{3a} 2a = \frac{\log 2a}{\log 3a}.$$

Thus, we get: $xy = \dfrac{\log a}{\log 2a} \cdot \dfrac{\log 2a}{\log 3a} = \dfrac{\log a}{\log 3a} = \log_{3a} a.$

So we can now try first, forming the exponents in the equality, $2^{1-xy} = 3^{y-xy}$.

$$1 - xy = \log_{3a} 3a - \log_{3a} a = \log_{3a} \tfrac{3a}{a} = \log_{3a} 3 \Rightarrow 3 = (3a)^{1-xy}.$$

$$y - xy = \log_{3a} 2a - 1 \log_{3a} a = \log_{3a} \tfrac{2a}{a} = \log_{3a} 2 \Rightarrow 2 = (3a)^{y-xy}.$$

Let's next, form each side in the equality.   Then, we get:

$$3 = (3a)^{1-xy} \Rightarrow 3^{y-xy} = \{(3a)^{1-xy}\}^{y-xy}.$$

$2 = (3a)^{y-xy} \Rightarrow 2^{1-xy} = \{(3a)^{y-xy}\}^{1-xy} = \{(3a)^{1-xy}\}^{y-xy}$, since we have an identity where $B^{mn} = B^{nm}$.

Thus, we can see that $2^{1-xy} = 3^{y-xy}$.

## Suggestions or Solutions
## To the Problem in the Example 2

**Assuming $3^x = 4^y = 6^z$, show that $\dfrac{1}{2y} = \dfrac{1}{z} - \dfrac{1}{x}$.**

Setting $3^x = 4^y = 6^z = k$ and applying logs to it, we get: $\log k = x \log 3 = y \log 4 = z \log 6$.

Then, $x = \dfrac{\log k}{\log 3}$, $y = \dfrac{\log k}{\log 4}$, and $z = \dfrac{\log k}{\log 6}$.

Thus, $\dfrac{1}{x} = \dfrac{\log 3}{\log k}$, $\dfrac{1}{2y} = \dfrac{\log 4}{2\log k} = \dfrac{\log 2}{\log k}$, and $\dfrac{1}{z} = \dfrac{\log 6}{\log k}$.

So $\dfrac{1}{z} - \dfrac{1}{x} = \dfrac{\log 6}{\log k} - \dfrac{\log 3}{\log k} = \dfrac{\log 6 - \log 3}{\log k} = \dfrac{\log \frac{6}{3}}{\log k} = \dfrac{\log 2}{\log k} = \dfrac{1}{2y}$.

Therefore, $\dfrac{1}{2y} = \dfrac{1}{z} - \dfrac{1}{x}$.

*If not quite sure of the idea behind the processes above, follow the steps below:*

Forming the equality, $\dfrac{1}{2y} = \dfrac{1}{z} - \dfrac{1}{x}$, which is to be proved, we need to get $x, y$, and $z$.

We can get those from the assumption. How then, do we get them?

Apply logs to the equality in the assumption.
Before taking logs of the terms in the equality however, we may want to set the equality equal to a constant so that we can readily extract $x, y$, and $z$.

Let's now, set: $3^x = 4^y = 6^z = k$, and apply logs to this equaliy.

Then, we get: $\log k = \log 3^x = \log 4^y = \log 6^z$.

Thus, we get: $\log k = x \log 3 = y \log 4 = z \log 6$.

So we get: $x = \dfrac{\log k}{\log 3}$, $y = \dfrac{\log k}{\log 4}$, and $z = \dfrac{\log k}{\log 6}$.

Let's next, form the terms in the equality to be proved.

Then, we get: $\dfrac{1}{x} = \dfrac{\log 3}{\log k}$, $2y = \dfrac{2\log k}{\log 4} \;\Rightarrow\; \dfrac{1}{2y} = \dfrac{\log 4}{2\log k}$, and $\dfrac{1}{z} = \dfrac{\log 6}{\log k}$.

Thus, we get: $\dfrac{1}{z} - \dfrac{1}{x} = \dfrac{\log 6}{\log k} - \dfrac{\log 3}{\log k} = \dfrac{\log 6 - \log 3}{\log k} = \dfrac{\log \frac{6}{3}}{\log k} = \dfrac{\log 2}{\log k} \;\Rightarrow\; \dfrac{1}{z} - \dfrac{1}{x} = \dfrac{\log 2}{\log k}$.

Meanwhile, $\dfrac{1}{2y} = \dfrac{\log 4}{2\log k} = \dfrac{\log 2^2}{2\log k} = \dfrac{2\log 2}{2\log k} = \dfrac{\log 2}{\log k}$.

Therefore, we can see that $\dfrac{1}{2y} = \dfrac{1}{z} - \dfrac{1}{x}$.

**In short:**

Setting $3^x = 4^y = 6^z = k$ and applying logs to it, we get: $\log k = x\log 3 = y\log 4 = z\log 6$.

Then, $x = \dfrac{\log k}{\log 3}$, $y = \dfrac{\log k}{\log 4}$, and $z = \dfrac{\log k}{\log 6}$.

Thus, $\dfrac{1}{x} = \dfrac{\log 3}{\log k}$, $\dfrac{1}{2y} = \dfrac{\log 4}{2\log k} = \dfrac{\log 2}{\log k}$, and $\dfrac{1}{z} = \dfrac{\log 6}{\log k}$.

So $\dfrac{1}{z} - \dfrac{1}{x} = \dfrac{\log 6}{\log k} - \dfrac{\log 3}{\log k} = \dfrac{\log 6 - \log 3}{\log k} = \dfrac{\log \frac{6}{3}}{\log k} = \dfrac{\log 2}{\log k} = \dfrac{1}{2y}$.

Therefore, $\dfrac{1}{2y} = \dfrac{1}{z} - \dfrac{1}{x}$.

## Examples A on Logarithms

0.   Assuming $11.2^a = 0.0112^b = 1000$, show that $\dfrac{1}{a} - \dfrac{1}{b} = 1$.

1.   Solve the equation as follows: $\log \frac{x-2}{2} = \dfrac{\log x - \log 2}{2}$.

2.   Suppose that $a$, $b$, and $c > 0$, but $\neq 1$. Suppose also, that $c = ab$, and that $a^x = b^y = c^z$. Then, show that $z = \dfrac{xy}{x+y}$.

3.   Assuming $9^{a+1} = 45$, $\log_7 3 = \frac{1}{b}$, and $\left(\frac{343}{45}\right)^t = 105$, find $t$, and then, put $t$ in terms of $a$ and $b$.

## Suggestions or Solutions
## To the Problem in the Example 0

**Assuming $11.2^a = 0.0112^b = 1000$, show that $\dfrac{1}{a} - \dfrac{1}{b} = 1$.**

We can put the problem in this way, too: $\mathbf{11.2^a = 0.0112^b = 1000} \Rightarrow \frac{1}{a} - \frac{1}{b} = 1$.

In other words, "By means of $\mathbf{11.2^a = 0.0112^b = 1000}$, show: $\frac{1}{a} - \frac{1}{b} = 1$."

So we get to show the equality $\frac{1}{a} - \frac{1}{b} = 1$ using a fact that $\mathbf{11.2^a = 0.0112^b = 1000}$. How?

We can extract $a$ and $b$ from the fact. Then, we put them into the right hand side of the equality, and do the algebra. How do we extract them, though?

Applying logs to the fact, we get:

$\mathbf{\log 11.2^a = \log 0.0112^b = \log 1000} \Rightarrow a \log 11.2 = b \log 0.0112 = \log 1000 = \log 10^3 = 3.$

So we get: $a = \dfrac{3}{\log 11.2} \Rightarrow \dfrac{1}{a} = \dfrac{\log 11.2}{3}$, and $b = \dfrac{3}{\log 0.0112} \Rightarrow \dfrac{1}{b} = \dfrac{\log 0.0112}{3}$.

Thus, $\dfrac{1}{a} - \dfrac{1}{b} = \dfrac{\log 11.2}{3} - \dfrac{\log 0.0112}{3} = \dfrac{(\log 11.2 - \log 0.0112)}{3} = \dfrac{\log \frac{11.2}{0.0112}}{3} = \dfrac{\log 1000}{3} = 1.$

Note that $\dfrac{1}{a} - \dfrac{1}{b} = 1 \Leftrightarrow b - a = ab$, where $a$ and $b \neq 0$.

Thus, we can show: $b - a = ab$ instead if it is easier.

Now, what if we have: $23^a = 0.0023^b = 10000$ instead of $11.2^a = 0.0112^b = 1000$?

We still get $\frac{1}{a} - \frac{1}{b} = 1$.

How about: $3.5^a = 350^b = 100$?

We still get: $\frac{1}{a} - \frac{1}{b} = 1$, too.

What then, if $12.5^a = 1.25^b = 10$?

We get: $\frac{1}{a} - \frac{1}{b} = 1$, still.  Noticed the pattern?

## Suggestions or Solutions
## To the Problem in the Example 1

**Solve the following equation:** $\log \frac{x-2}{2} = \dfrac{\log x - \log 2}{2}$.

$\log \frac{x-2}{2} = \frac{\log x - \log 2}{2} = \frac{1}{2}\log \frac{x}{2} \Rightarrow 2\log \frac{x-2}{2} = \log \frac{x}{2} \Rightarrow \left(\frac{x-2}{2}\right)^2 = \frac{x}{2}$

$\Rightarrow (x-2)^2 = 2x \Rightarrow x^2 - 4x + 4 = 2x \Rightarrow x^2 - 6x + 4 = 0 \Rightarrow x = 3 \pm \sqrt{5}.$

Both $x = 3 + \sqrt{5}$ and $x = 3 - \sqrt{5} > 0$, so both are allowed for $\log x$.

$x = 3 - \sqrt{5} \Rightarrow \frac{x-2}{2} = \frac{3-\sqrt{5}-2}{2} = \frac{1-\sqrt{5}}{2} < 0$, so $x = 3 - \sqrt{5}$ is not allowed for $\log \frac{x-2}{2}$.

$x = 3 + \sqrt{5} \Rightarrow \frac{x-2}{2} = \frac{3+\sqrt{5}-2}{2} = \frac{1+\sqrt{5}}{2} > 0$, so $x = 3 + \sqrt{5}$ is allowed for $\log \frac{x-2}{2}$.

Therefore, $x = 3 + \sqrt{5}$.

*If not quite sure of the idea behind the processes above, follow the steps below:*

Such an equation as above can be called a log equation.

Hardly, we can directly solve log equations.

Normally, we remove log signs first, or begin with a substitution.
In this problem, we can remove log signs first.

Removing log signs in this case, we can get an ordinary algebraic equation for *x*.

$\log \frac{x-2}{2} = \frac{\log x - \log 2}{2} \Rightarrow 2\log \frac{x-2}{2} = \log x - \log 2 \Rightarrow \log(\frac{x-2}{2})^2 = \log \frac{x}{2} \Rightarrow \left(\frac{x-2}{2}\right)^2 = \frac{x}{2}.$

Of course, we can put it this way, too:

$\dfrac{\log x - \log 2}{2} = \frac{1}{2}(\log x - \log 2) = \frac{1}{2}\log \frac{x}{2} = \log \sqrt{\frac{x}{2}}.$

Thus, we get:

$$\log\frac{x-2}{2} = \frac{\log x - \log 2}{2} = \log\sqrt{\frac{x}{2}} \Rightarrow \frac{x-2}{2} = \sqrt{\frac{x}{2}} \Rightarrow \frac{(x-2)^2}{4} = \frac{x}{2}.$$

So the last equation above is just a quadratic equation, yet after solving the equation, we want to check to see if all the roots are OK.

That is because negative cases can get involved due to the squaring operation done at the stage where $\frac{x-2}{2} = \sqrt{\frac{x}{2}}$.

Expanding the left hand side in $\frac{(x-2)^2}{4} = \frac{x}{2}$, we get: $\frac{x^2 - 4x + 4}{4}$.

Thus, we get: $\frac{x^2 - 4x + 4}{4} = \frac{x}{2} \Rightarrow x^2 - 4x + 4 = 2x \Rightarrow x^2 - 6x + 4 = 0$.

The left hand side doesn't simply get factorized to a form of $(x - a)(x - b)$.

So using the quadratic formula, we get: $x = 3 \pm \sqrt{5}$.

Since both $3 + \sqrt{5}$ and $3 - \sqrt{5} > 0$, they have no problem with $\log x$. However, we want to make sure if both are fine with $\log\frac{x-2}{2}$, too.

$x = 3 + \sqrt{5} \Rightarrow \log\frac{x-2}{2} = \log\frac{3+\sqrt{5}-2}{2} = \log\frac{1+\sqrt{5}}{2}$, which means it's OK, because $\frac{1+\sqrt{5}}{2} > 0$.

$x = 3 - \sqrt{5} \Rightarrow \log\frac{x-2}{2} = \log\frac{3-\sqrt{5}-2}{2} = \log\frac{1-\sqrt{5}}{2}$, which means it's not OK, because $\frac{1-\sqrt{5}}{2} < 0$.

Therefore, the solution is $x = 3 + \sqrt{5}$.

## Suggestions or Solutions
## To the Problem in the Example 2

**Suppose that $a$, $b$, and $c > 0$, but $\neq 1$. Suppose also, that $c = ab$, and that $a^x = b^y = c^z$.**

**Then, show that $z = \dfrac{xy}{x+y}$.**

Setting first: $a^x = b^y = c^z = k$, we get:

$\log a^x = \log b^y = \log c^z = \log k \Rightarrow x \log a = y \log b = z \log c = \log k$.

Thus, we get: $x = \dfrac{\log k}{\log a}$, $y = \dfrac{\log k}{\log b}$, and $z = \dfrac{\log k}{\log c}$.

So we get: $xy = \dfrac{\log k}{\log a} \cdot \dfrac{\log k}{\log b} = \dfrac{(\log k)^2}{(\log a)(\log b)}$.

And $x + y = \dfrac{\log k}{\log a} + \dfrac{\log k}{\log b} = \log k \dfrac{(\log b + \log a)}{(\log a)(\log b)} = \dfrac{(\log k)(\log ab)}{(\log a)(\log b)}$.

Thus, we get: $\dfrac{xy}{x+y} = \dfrac{(\log k)^2}{(\log a)(\log b)} \cdot \dfrac{(\log a)(\log b)}{(\log k)(\log ab)} = \dfrac{\log k}{\log ab}$.

And we have: $z = \dfrac{\log k}{\log c}$, and $c = ab$, too.

Thus, we get: $z = \dfrac{\log k}{\log c} = \dfrac{\log k}{\log ab}$, so we get: $z = \dfrac{xy}{x+y}$.

*If not quite sure of the idea behind the processes above, follow the steps below:*

What we want to show is of course, under the assumption that $a$, $b$, and $c > 0$, but $\neq 1$.

And we get to show the equality $z = \frac{xy}{x+y}$ using a fact that $a^x = b^y = c^z$ where $c = ab$. How?

We can put each of $x$, $y$, and $z$ in terms of $a$, $b$, and $c$. Then, we put each of them into the equality, and do the algebra. How do we get $a$, $b$, and $c$, though?

We don't actually get $a$, $b$, and $c$, themselves. What then, do we get?

We can extract logs of $a$, $b$, and $c$ from the fact given, which is the equality $a^x = b^y = c^z$, then we put each of $x$, $y$, and $z$ in terms of the logs. So let's extract them now.

We may want to first, set: $a^x = b^y = c^z = k$ so that we can readily refer to the fact. Then, we can extract them applying logs to the fact. So applying logs to the fact, we get:

$$\log a^x = \log b^y = \log c^z = \log k \Rightarrow x \log a = y \log b = z \log c = \log k.$$

Thus, we get: $x = \dfrac{\log k}{\log a}$, $y = \dfrac{\log k}{\log b}$, and $z = \dfrac{\log k}{\log c}$.

Now, we can put each of $x$, $y$, and $z$ into the equality where $z = \frac{xy}{x+y}$.

To begin with, we get: $xy = \dfrac{\log k}{\log a} \cdot \dfrac{\log k}{\log b} = \dfrac{(\log k)^2}{(\log a)(\log b)}$.

Next, we get:

$$x + y = \dfrac{\log k}{\log a} + \dfrac{\log k}{\log b} = \log k \left( \dfrac{1}{\log a} + \dfrac{1}{\log b} \right) = \log k \dfrac{(\log b + \log a)}{(\log a)(\log b)} = \dfrac{(\log k)(\log ab)}{(\log a)(\log b)}.$$

Thus, we get: $\dfrac{xy}{x+y} = xy \dfrac{1}{x+y} = \dfrac{(\log k)^2}{(\log a)(\log b)} \cdot \dfrac{(\log a)(\log b)}{(\log k)(\log ab)} = \dfrac{\log k}{\log ab}$.

And we have: $z = \dfrac{\log k}{\log c}$, and $c = ab$, too.

Thus, we get: $z = \dfrac{\log k}{\log c} = \dfrac{\log k}{\log ab}$, so we can see that $z = \dfrac{xy}{x+y}$.

**In short:**

Setting first: $a^x = b^y = c^z = k$, we get:

$\log a^x = \log b^y = \log c^z = \log k \Rightarrow x \log a = y \log b = z \log c = \log k$.

Thus, we get: $x = \dfrac{\log k}{\log a}$, $y = \dfrac{\log k}{\log b}$, and $z = \dfrac{\log k}{\log c}$.

So we get: $xy = \dfrac{\log k}{\log a} \cdot \dfrac{\log k}{\log b} = \dfrac{(\log k)^2}{(\log a)(\log b)}$.

And $x + y = \dfrac{\log k}{\log a} + \dfrac{\log k}{\log b} = \log k \dfrac{(\log b + \log a)}{(\log a)(\log b)} = \dfrac{(\log k)(\log ab)}{(\log a)(\log b)}$.

Thus, we get: $\dfrac{xy}{x+y} = \dfrac{(\log k)^2}{(\log a)(\log b)} \cdot \dfrac{(\log a)(\log b)}{(\log k)(\log ab)} = \dfrac{\log k}{\log ab}$.

And we have: $z = \dfrac{\log k}{\log c}$, and $c = ab$, too.

Thus, we get: $z = \dfrac{\log k}{\log c} = \dfrac{\log k}{\log ab}$, so we get: $z = \dfrac{xy}{x+y}$.

## Suggestions or Solutions
## To the Problem in the Example 3

Assuming $9^{a+1} = 45$, $\log_7 3 = \frac{1}{b}$, and $\left(\frac{343}{45}\right)^t = 105$, **find** $t$, **and then, put** $t$ **in terms of** $a$ **and** $b$.

$$\log\left(\tfrac{343}{45}\right)^t = \log 105 \Rightarrow t \log \tfrac{343}{45} = \log 105 \Rightarrow t = \frac{\log 105}{\log \tfrac{343}{45}} \Rightarrow t = \frac{\log 105}{\log 343 - \log 45}.$$

$$9^{a+1} = 45 \Rightarrow \log 9^{a+1} = \log 45 \Rightarrow (a+1)\log 9 = \log 5 + \log 9 \Rightarrow a + 1 = \frac{\log 5 + \log 9}{\log 9}$$

$$\Rightarrow a = \frac{\log 5}{\log 9} = \frac{\log 5}{2\log 3} \Rightarrow 2a = \frac{\log 5}{\log 3} = \log_3 5. \quad \text{And } \log_7 3 = \tfrac{1}{b} \Rightarrow \tfrac{1}{\log_7 3} = \log_3 7.$$

And we have $105 = 3 \cdot 5 \cdot 7$, $343 = 7^3$, and $45 = 3^2 5$. So we get:

$\log 105 = \log (3 \cdot 5 \cdot 7) = \log 3 + \log 5 + \log 7$, $\log 343 = \log 7^3 = 3 \log 7$, and

$\log 45 = \log 5 + 2 \log 3$.

Thus, we get: $t = \dfrac{\log 105}{\log 343 - \log 45} = \dfrac{\log 3 + \log 5 + \log 7}{3\log 7 - \log 5 - 2\log 3} = \dfrac{\log_3 3 + \log_3 5 + \log_3 7}{3\log_3 7 - \log_3 5 - 2\log_3 3}$

$$= \frac{1 + \log_3 5 + \log_3 7}{3\log_3 7 - \log_3 5 - 2} = \frac{1 + 2a + b}{3b - 2a - 2}.$$

*If not quite sure of the idea behind the processes above, follow the steps below:*

We can readily find $t$. Finding $t$ though, is just one thing, and putting it in terms of $a$ and $b$ is another.

Applying logs to the equation for $t$ where $\left(\frac{343}{45}\right)^t = 105$, we get:

$$\log\left(\tfrac{343}{45}\right)^t = \log 105 \Rightarrow t \log \tfrac{343}{45} = \log 105 \Rightarrow t = \frac{\log 105}{\log \frac{343}{45}} \Rightarrow t = \frac{\log 105}{\log 343 - \log 45}.$$

And of course, by the definition for logs, we can get: $t = \log_{\frac{343}{45}} 105$, which is however, no other than $\dfrac{\log 105}{\log 343 - \log 45}$, because we have: $\log_{\frac{343}{45}} 105 = \dfrac{\log 105}{\log \frac{343}{45}}$.

Now, we need to put $t$ in terms of $a$ and $b$, so we may want to get $a$ and $b$, first. How?

We are given two facts, one is an equation for $a$, and the other is an equation for $b$.

So solving the equations for $a$ and $b$ respectively, we can get $a$ and $b$.

Thus, beginning with $a$, we get: $9^{a+1} = 45 \Rightarrow \log 9^{a+1} = \log 45$

$\Rightarrow (a+1)\log 9 = \log 5 + \log 9 \Rightarrow a+1 = \frac{\log 5 + \log 9}{\log 9} \Rightarrow a = \frac{\log 5 + \log 9}{\log 9} - 1 = \frac{\log 5}{\log 9} \Rightarrow a = \frac{\log 5}{\log 9}$.

Next, moving on to $b$, we get: $\log_7 3 = \frac{1}{b} \Rightarrow b = \frac{1}{\log_7 3} = \log_3 7$, where the base is 3.

So we may want to put $a$ in this way: $a = \frac{\log 5}{\log 9} = \frac{\log 5}{\log 3^2} = \frac{\log 5}{2\log 3} \Rightarrow 2a = \frac{\log 5}{\log 3} = \log_3 5$.

Of course, we can put it this way, too: $a = \dfrac{\log 5}{\log 9} = \dfrac{\log_3 5}{\log_3 9} = \dfrac{\log_3 5}{2} \Rightarrow 2a = \log_3 5$.

How come we get: $\frac{\log 5}{\log 9} = \frac{\log_3 5}{\log_3 9}$, though?

We will see how to get it shortly. So let's see for now, what we can do about *t*.

We have: $t = \dfrac{\log 105}{\log 343 - \log 45}$.

So it is made of logs, where the antilogs are 105, 343, and 45.
Thus, we may want to see what the antilogs are composed of.    How though?

Factorizing the antilogs, we can readily see the components.

We have: $105 = 3{\cdot}5{\cdot}7$, $343 = 7^3$, and $45 = 3^2 5$.    So we can see that:

$\log 105 = \log (3{\cdot}5{\cdot}7) = \log 3 + \log 5 + \log 7$, $\log 343 = \log 7^3 = 3 \log 7$, and

$\log 45 = \log 5 + 2 \log 3$.

Thus, we get: $t = \dfrac{\log 105}{\log 343 - \log 45} = \dfrac{\log 3 + \log 5 + \log 7}{3 \log 7 - (\log 5 + 2 \log 3)} = \dfrac{\log 3 + \log 5 + \log 7}{3 \log 7 - \log 5 - 2 \log 3}$.

Now, we have: $t = \dfrac{\log 3 + \log 5 + \log 7}{3 \log 7 - \log 5 - 2 \log 3}$, $2a = \log_3 5$, and $b = \log_3 7$.

Besides, we have: $\dfrac{\log x}{\log y} = \dfrac{\frac{\log x}{\log z}}{\frac{\log y}{\log z}} = \dfrac{\log_z x}{\log_z y}$, also. (So we get: $\dfrac{\log 5}{\log 9} = \dfrac{\log_3 5}{\log_3 9}$, too.)

Thus, dividing by $\log 3$ both the numerator and denominator in $\dfrac{\log 3 + \log 5 + \log 7}{3 \log 7 - \log 5 - 2 \log 3}$,

we get: $t = \dfrac{\log_3 3 + \log_3 5 + \log_3 7}{3 \log_3 7 - \log_3 5 - 2 \log_3 3} = \dfrac{1 + \log_3 5 + \log_3 7}{3 \log_3 7 - \log_3 5 - 2}$.

Thus, we get: $t = \dfrac{1 + \log_3 5 + \log_3 7}{3 \log_3 7 - \log_3 5 - 2} = \dfrac{1 + 2a + b}{3b - 2a - 2}$.

**In short:**

$$\log\left(\tfrac{343}{45}\right)^t = \log 105 \Rightarrow t \log \tfrac{343}{45} = \log 105 \Rightarrow t = \frac{\log 105}{\log \tfrac{343}{45}} \Rightarrow t = \frac{\log 105}{\log 343 - \log 45}.$$

$$9^{a+1} = 45 \Rightarrow \log 9^{a+1} = \log 45 \Rightarrow (a+1)\log 9 = \log 5 + \log 9 \Rightarrow a+1 = \frac{\log 5 + \log 9}{\log 9}$$

$$\Rightarrow a = \frac{\log 5}{\log 9} = \frac{\log 5}{2 \log 3} \Rightarrow 2a = \frac{\log 5}{\log 3} = \log_3 5. \quad \text{And } \log_7 3 = \tfrac{1}{b} \Rightarrow \tfrac{1}{\log_7 3} = \log_3 7.$$

And we have $105 = 3 \cdot 5 \cdot 7$, $343 = 7^3$, and $45 = 3^2 5$. So we get:

$\log 105 = \log (3 \cdot 5 \cdot 7) = \log 3 + \log 5 + \log 7$, $\log 343 = \log 7^3 = 3 \log 7$, and

$\log 45 = \log 5 + 2 \log 3$.

Thus, we get: $t = \dfrac{\log 105}{\log 343 - \log 45} = \dfrac{\log 3 + \log 5 + \log 7}{3 \log 7 - \log 5 - 2 \log 3} = \dfrac{\log_3 3 + \log_3 5 + \log_3 7}{3 \log_3 7 - \log_3 5 - 2 \log_3 3}$

$$= \frac{1 + \log_3 5 + \log_3 7}{3 \log_3 7 - \log_3 5 - 2} = \frac{1 + 2a + b}{3b - 2a - 2}.$$

## Examples B on Logarithms

0.0.  Assuming that $(\log xy)^2 = (\log x)(\log y) + \log x - \log y - 1$, find the value of $xy$.

0.1.  Assuming that $ab$, $bc$, and $ca > 0$, and that $(\log ac)(\log bc) + 1 = 0$ is an equation for $c$, find an additional condition on $a$ and $b$ so that the equation has real roots.

1.  Suppose an equation for $x$, $(\log 2x)(\log 3x) = -a^2$ has two real roots. Then:

1.0.  Find the extent of $a$.

1.1.  Find the value of the product of the roots.

2.  Solve the following equation for $x$:

$$\left(\frac{x}{10^{100}}\right)^u = \frac{\log 10^x}{x} \text{ where } u = (\log x^v - \log x^7 + 10)^3 \text{ where } v = \log x.$$

## Suggestions or Solutions
## To the Problem 0 in the Example 0

**Assuming that $(\log xy)^2 = (\log x)(\log y) + \log x - \log y - 1$, find the value of $xy$.**

Setting $u = \log x$ and $v = \log y$, we get:

$(\log xy)^2 = (\log x + \log y)^2 = (\log x)(\log y) + \log x - \log y - 1$

$\Rightarrow (u + v)^2 = uv + u - v - 1 \Rightarrow u^2 + 2uv + v^2 = uv + u - v - 1$

$\Rightarrow u^2 + uv - u + v + v^2 + 1 = 0.$

Taking the final equation above as an equation for $u$, we can get a quadratic equation as follows: $u^2 + (v - 1)u + v^2 + v + 1 = 0.$

Since the equaion above needs to have real roots, we need to have:
$D = (v - 1)^2 - 4(v^2 + v + 1) \geq 0.$

And simplifying the expression above, and rewriting the result, we can get:

$D = v^2 - 2v + 1 - 4v^2 - 4v - 4 = -3v^2 - 6v - 3 = -3(v^2 + 2v + 1) = -3(v + 1)^2 \geq 0$

$\Rightarrow v = -1 \Rightarrow \log y = -1.$

Thus, $y = 10^{-1}.$

Next, $v = -1 \Rightarrow u^2 + (v - 1)u + v^2 + v + 1 = u^2 - 2u + 1 = (u - 1)^2 = 0.$

So $u = 1$, and thus, $\log x = 1 \Rightarrow x = 10.$   Therefore, $xy = 1.$

*If not quite sure of the idea behind the processes above, follow the steps below:*

We have two unknowns, which are $x$ and $y$, so we need to have two equations, but are given one equation only.   It's OK, though, in this case.   How come?

The equation given is quadratic. Working with a quadratic equation, we often take from the equation a special value or expression, which is intrinsic to an equation quadratic.

And the value or expression is called the discriminant, usually denoted by $D$.

Suppose that we have an equation for $x$ as follows: $ax^2 + bx + c = 0$ where $a$, $b$, and $c$ are constant. Then, the discriminant $D = b^2 - 4ac$. If in particular, $b = 2k$, that is, $b$ is even, we often set: $D/4 = k - ac$.     What then, is the discriminant about?

It is in fact, what's inside the square root sign in the quadratic formula, which is below:

$x = \dfrac{-b \pm \sqrt{b^2 - 4ac}}{2a}$. Thus, the formula can be put this way, too: $x = \dfrac{-b \pm \sqrt{D}}{2a}$, for short.

And if $b = 2k$, we can put it this way, too: $x = \dfrac{-k \pm \sqrt{k^2 - ac}}{a}$. That's because we get:

$$x = \frac{-b \pm \sqrt{b^2 - 4ac}}{2a} = \frac{-2k \pm \sqrt{4k^2 - 4ac}}{2a} = \frac{-2k \pm 2\sqrt{k^2 - ac}}{2a} = \frac{-k \pm \sqrt{k^2 - ac}}{a}.$$

And we know what's inside the square root sign is $\geq 0$ if the root is real.
Therefore, if a quadratic equation has real roots, the discriminant $D$ has to be $\geq 0$.

If $D = 0$, the equation has a double root, which is: $\dfrac{-b}{2a}$.

If $D < 0$, the equation has no real root.

And if $D > 0$, the equation has two different real roots, which are: $\dfrac{-b \pm \sqrt{b^2 - 4ac}}{2a}$.

Since we have a log equation however, we want to check what's inside the log, too.

That is, we want to consider the requirement on logarithms as follows:

• The *antilog* and *base* in a log have to be *positive*, but the *base cannot* be 1.

Now, rewriting the equation given, we can have better look at the equation.
And rewriting it, we can begin with, for simplicity, setting: $u = \log x$, and $v = \log y$.

Then, we get: $(\log xy)^2 = (\log x + \log y)^2 = (\log x)(\log y) + \log x - \log y - 1$

$\Rightarrow (u + v)^2 = uv + u - v - 1 \Rightarrow u^2 + 2uv + v^2 = uv + u - v - 1$

$\Rightarrow u^2 + uv - u + v + v^2 + 1 = 0.$   What then?

We can now take the last equation above as an equation for $u$, and put it the way below:

$u^2 + (v - 1)u + v^2 + v + 1 = 0$, which is a quadratic equation for $u$.   So?

If the equation has the solution, the solution has to be real.
That is, the equation needs to have a real root.
And since the equation above needs to have a real root, the discriminant $D$ has to be $\geq 0$.

That is, we need: $D = (v - 1)^2 - 4 \cdot 1 \cdot (v^2 + v + 1) \geq 0$.

Meanwhile:
$D = v^2 - 2v + 1 - 4v^2 - 4v - 4 = -3v^2 - 6v - 3 = -3(v^2 + 2v + 1) = -3(v + 1)^2$.

So unless $v = -1$, we get: $D = -3(v + 1)^2 < 0$, that is, $D < 0$, which is not what we want.
What then, do we need to have?


We need to have: $v = \log y = -1$, since we need to have: $D \geq 0$.

So by the definition for logs, we get: $y = 10^{-1} = \frac{1}{10} = 0.1$.   How come?



We know: $\log y = \log_{10} y$, and the definition for logs is: $A = b^x \Leftrightarrow x = \log_b A$.

So by the definition for logs, we can get: $y = 10^{-1} \Leftrightarrow -1 = \log_{10} y = \log y = \log 10^{-1}$.
What then, is the next?

We want to get the value of $u$ so that we can get the value of $x$, since we set: $u = \log x$.

So next, going back to the equation $u^2 + (v-1)u + v^2 + v + 1 = 0$, since $v = -1$, we get:

$u^2 + (v-1)u + v^2 + v + 1 = u^2 - 2u + 1 = (u-1)^2 = 0$.

So we get: $u = 1$, and thus, we get: $\log x = 1 \Rightarrow x = 10$.

Therefore, $xy = 1$.

**In short:**

Setting $u = \log x$ and $v = \log y$, we get:

$(\log xy)^2 = (\log x + \log y)^2 = (\log x)(\log y) + \log x - \log y - 1$

$\Rightarrow (u + v)^2 = uv + u - v - 1 \Rightarrow u^2 + 2uv + v^2 = uv + u - v - 1$

$\Rightarrow u^2 + uv - u + v + v^2 + 1 = 0$.

Taking the final equation above as an equation for $u$, we can get a quadratic equation as follows: $u^2 + (v-1)u + v^2 + v + 1 = 0$.

Since the equaion above needs to have real roots, we need to have:
$D = (v-1)^2 - 4(v^2 + v + 1) \geq 0$.

Expanding the expression above, we get:

$D = v^2 - 2v + 1 - 4v^2 - 4v - 4 = -3v^2 - 6v - 3 = -3(v^2 + 2v + 1) = -3(v+1)^2 \geq 0$

$\Rightarrow v = -1 \Rightarrow \log y = -1$.

Thus, $y = 10^{-1}$.

Next, $v = -1 \Rightarrow u^2 + (v-1)u + v^2 + v + 1 = u^2 - 2u + 1 = (u-1)^2 = 0$.

So $u = 1$, and thus, $\log x = 1 \Rightarrow x = 10$. Therefore, $xy = 1$.

## Suggestions or Solutions
## To the Problem 1 in the Example 0

**Assuming that $ab$, $bc$, and $ca > 0$, and that $(\log ac)(\log bc) + 1 = 0$ is an equation for $c$, find an additional condition on $a$ and $b$ so that the equation has real roots.**

$(\log ac)(\log bc) + 1 = (\log a + \log c)(\log b + \log c) + 1$.

Setting $u = \log a$, $v = \log b$, and $w = \log c$, we get:
$(u + w)(v + w) + 1 = w^2 + (u + v)w + uv + 1 = 0$.

Since the equation above needs to have real roots, the discriminant $D \geq 0$.
Taking thus, the discriminant $D$, and setting it to be $\geq 0$, we get:
$D = (u + v)^2 - 4(uv + 1) = u^2 + 2uv + v^2 - 4uv - 4 = u^2 - 2uv + v^2 - 4 = (u - v)^2 - 4 \geq 0$.

So we get: $(u - v)^2 \geq 4 \Rightarrow u - v \leq -2$, or $u - v \geq 2$.

We have: $u - v = \log a - \log b = \log \frac{a}{b}$. Thus, we get: $\log \frac{a}{b} \leq -2$, or $\log \frac{a}{b} \geq 2$.

Therefore (by the definition for logs), we get: $\frac{a}{b} \leq 10^{-2}$, or $\frac{a}{b} \geq 10^2$.

*If not quite sure of the idea behind the processes above, follow the steps below:*

To begin with, what do we mean by an equation with real roots?

The equation is probably quadratic, so it is very likely that we work with its discriminant. How can we see though, it's a quadratic equation?

We can begin with expanding the left hand side of the equation given so that we can see better the unknown $c$.

Then, we get: **(log *ac*)(log *bc*) + 1 = (log *a* + log *c*)(log *b* + log *c*) + 1 = 0.**

And next, since the equation above is an equation for *c*, taking *a* and *b* as constants, and taking **log *c*** as the unknown, we can take the equation as a quadratic equation for **log *c*.**

Setting in fact: *u* = log *a*, *v* = log *b*, and *w* = log *c*, we get a quadratic equation for *w* as follows: $(u + w)(v + w) + 1 = w^2 + (u + v)w + uv + 1 = 0.$

Since the equation above needs to have real roots, the discriminant **$D \geq 0$**. So we get:

$$D = (u + v)^2 - 4(uv + 1) = u^2 + 2uv + v^2 - 4uv - 4 = u^2 - 2uv + v^2 - 4 = (u - v)^2 - 4 \geq 0.$$

Thus, we get: $(u - v)^2 \geq 4 \Rightarrow u - v \leq -2$, or $u - v \geq 2$.

Now, since we have: *u* = **log *a***, and *v* = **log *b***, we get: $u - v = \log a - \log b = \log \frac{a}{b}$.

So we get: $\log \frac{a}{b} \leq -2$, or $\log \frac{a}{b} \geq 2$. Therefore, we get: $\frac{a}{b} \leq 10^{-2}$, or $\frac{a}{b} \geq 10^2$.

We can put it this way, too: **$100a \leq b$, or $a \geq 100b$.**


**In short:**

**(log *ac*)(log *bc*) + 1 = (log *a* + log *c*)(log *b* + log *c*) + 1.**

Setting *u* = log *a,* *v* = log *b*, and *w* = log *c*, we get:
$(u + w)(v + w) + 1 = w^2 + (u + v)w + uv + 1 = 0.$

Since the equation above needs to have real roots, the discriminant **$D \geq 0$.**
Taking thus, the discriminant **$D$**, and setting it to be $\geq 0$, we get:
$$D = (u + v)^2 - 4(uv + 1) = u^2 + 2uv + v^2 - 4uv - 4 = u^2 - 2uv + v^2 - 4 = (u - v)^2 - 4 \geq 0.$$

So we get: $(u - v)^2 \geq 4 \Rightarrow u - v \leq -2$, or $u - v \geq 2$.

We have: $u - v = \log a - \log b = \log \frac{a}{b}$. Thus, we get: $\log \frac{a}{b} \leq -2$, or $\log \frac{a}{b} \geq 2$.

Therefore (by the definition for logs), we get: $\frac{a}{b} \leq 10^{-2}$, or $\frac{a}{b} \geq 10^2$.

## Suggestions or Solutions
## To the Problem 0 in the Example 1

**Assuming an equation for $x$, $(\log 2x)(\log 3x) = -a^2$ has two real roots, find the extent of $a$.**

$(\log 2x)(\log 3x) = (\log 2 + \log x)(\log 3 + \log x)$

$= (\log x)^2 + (\log 2 + \log 3)(\log x) + (\log 2)(\log 3) = -a^2.$

Thus, setting $u = \log x$, we get: $u^2 + (\log 6)u + (\log 2)(\log 3) + a^2 = 0.$

The equation above is a quadratic for $u$, and has two real roots, so the discriminant $D > 0$.

So we get: $D = (\log 6)^2 - 4(\log 2)(\log 3) - 4a^2 > 0.$

Meanwhile, $\log 6 = \log 3 + \log 2 \Rightarrow (\log 6)^2 = (\log 3)^2 + 2(\log 3)(\log 2) + (\log 2)^2.$

Thus, $D = (\log 6)^2 - 4(\log 2)(\log 3) - 4a^2 = (\log 3)^2 - 2(\log 3)(\log 2) + (\log 2)^2 - 4a^2$

$= (\log 3 - \log 2)^2 - 4a^2 = (\log \tfrac{3}{2})^2 - 4a^2 = (\log \tfrac{3}{2} + 2a)(\log \tfrac{3}{2} - 2a) > 0.$

Therefore, $-\log \tfrac{3}{2} < 2a < \log \tfrac{3}{2} \Rightarrow -\tfrac{1}{2}\log \tfrac{3}{2} < a < \tfrac{1}{2}\log \tfrac{3}{2} \Rightarrow \log \sqrt{\tfrac{2}{3}} < a < \log \sqrt{\tfrac{3}{2}}.$

*If not quite sure of the idea behind the processes above, follow the steps below:*

The equation given is similar in structure to the one in the previous example.
So we can expect that the equation is quadratic with respect to **$\log x$**.

So let's begin with expanding the left hand side so that we can expose the equation.

$(\log 2x)(\log 3x) = (\log 2 + \log x)(\log 3 + \log x)$

$= (\log x)^2 + (\log 2 + \log 3)(\log x) + (\log 2)(\log 3) = -a^2.$

Next, setting: $u = \log x$, we get: $u^2 + (\log 6)u + (\log 2)(\log 3) + a^2 = 0$.

Now, the equation is a quadratic for $u$, and has to have two real roots, so the discriminant $D$ is not $\geq 0$ but $> 0$.

So we get: $D = (\log 6)^2 - 4 \cdot 1 \cdot \{(\log 2)(\log 3) + a^2\} = (\log 6)^2 - 4(\log 2)(\log 3) - 4a^2 > 0$.

Now, we could fill in the answer sheet the way as follows:

$(\log 6)^2 - 4(\log 2)(\log 3) - 4a^2 > 0 \Rightarrow 4a^2 < (\log 6)^2 - 4(\log 2)(\log 3)$

$$\Rightarrow a^2 < \frac{(\log 6)^2 - 4(\log 2)(\log 3)}{4}$$

$$\Rightarrow -\frac{\sqrt{(\log 6)^2 - 4(\log 2)(\log 3)}}{2} < a < \frac{\sqrt{(\log 6)^2 - 4(\log 2)(\log 3)}}{2}.$$

What if however, the test is multiple-choice, and the last expression above doesn't show up in the list of choices, but the list has $-\log\sqrt{\frac{2}{3}} < a < \log\sqrt{\frac{2}{3}}$, instead?

Notice that $\log 6 = \log(3 \cdot 2) = \log 3 + \log 2$.

So we get: $(\log 6)^2 = (\log 3 + \log 2)^2 = (\log 3)^2 + 2(\log 3)(\log 2) + (\log 2)^2$.

Thus, $D = (\log 6)^2 - 4(\log 2)(\log 3) - 4a^2 = (\log 3)^2 - 2(\log 3)(\log 2) + (\log 2)^2 - 4a^2$

$= (\log 3 - \log 2)^2 - 4a^2 = (\log\frac{3}{2})^2 - 4a^2 = (\log\frac{3}{2} + 2a)(\log\frac{3}{2} - 2a) > 0$.

So we get: $-\log\frac{3}{2} < 2a < \log\frac{3}{2} \Rightarrow -\frac{1}{2}\log\frac{3}{2} < a < \frac{1}{2}\log\frac{3}{2} \Rightarrow -\log(\frac{3}{2})^{\frac{1}{2}} < a < \frac{1}{2}\log(\frac{3}{2})^{\frac{1}{2}}$.

Therefore, we can see that: $-\log\sqrt{\frac{3}{2}} < a < \log\sqrt{\frac{3}{2}}$.

Besides, we can put it this way, too: $\log\sqrt{\frac{2}{3}} < a < \log\sqrt{\frac{3}{2}}$.

That's because $-\log\sqrt{\frac{3}{2}} = \log(\frac{3}{2})^{-\frac{1}{2}} = \log(\frac{2}{3})^{\frac{1}{2}} = \log\sqrt{\frac{2}{3}}$.

**In short:**

$(\log 2x)(\log 3x) = (\log 2 + \log x)(\log 3 + \log x)$

$= (\log x)^2 + (\log 2 + \log 3)(\log x) + (\log 2)(\log 3) = -a^2$.

Thus, setting $u = \log x$, we get: $u^2 + (\log 6)u + (\log 2)(\log 3) + a^2 = 0$.

The equation above is a quadratic for $u$, and has two real roots, so the discriminant $D > 0$.

So we get: $D = (\log 6)^2 - 4(\log 2)(\log 3) - 4a^2 > 0$.

Meanwhile, $\log 6 = \log 3 + \log 2 \Rightarrow (\log 6)^2 = (\log 3)^2 + 2(\log 3)(\log 2) + (\log 2)^2$.

Thus, $D = (\log 6)^2 - 4(\log 2)(\log 3) - 4a^2 = (\log 3)^2 - 2(\log 3)(\log 2) + (\log 2)^2 - 4a^2$

$= (\log 3 - \log 2)^2 - 4a^2 = (\log\frac{3}{2})^2 - 4a^2 = (\log\frac{3}{2} + 2a)(\log\frac{3}{2} - 2a) > 0$.

Therefore, we get:

$-\log\frac{3}{2} < 2a < \log\frac{3}{2} \Rightarrow -\frac{1}{2}\log\frac{3}{2} < a < \frac{1}{2}\log\frac{3}{2} \Rightarrow \log\sqrt{\frac{2}{3}} < a < \log\sqrt{\frac{3}{2}}$.

## Suggestions or Solutions
## To the Problem 1 in the Example 1

**Assuming an equation for $x$, $(\log 2x)(\log 3x) = -a^2$ has two real roots, find the value of the product of the roots.**

Setting $u = \log x$, we get: $u^2 + (\log 6)u + (\log 2)(\log 3) + a^2 = 0$.

Suppose that $s$ and $t$ are the two roots for the equation above.

Then, $(u - s)(u + t) = 0 \Rightarrow u^2 - (s + t)u + st = 0$. Thus, $s + t = -\log 6 = \log 6^{-1}$.

We have: $u = \log x$, too.
So we get: $\log x = s$ or $t \Rightarrow x = 10^s$ or $10^t$.

Therefore, the product is $10^s 10^t = 10^{s+t} = 10^{\log 6^{-1}} = 6^{-1} = \frac{1}{6}$.

*If not quite sure of the idea behind the processes above, follow the steps below:*

Setting: $u = \log x$, we get: $u^2 + (\log 6)u + (\log 2)(\log 3) + a^2 = 0$.

We don't even know however, the values of the roots for the equation above.

We can't get the actual values of the roots unless the value of $a$ is chosen. In fact, the actual values of the true roots are the values of $x$ that can satisfy the original equation.

Thus, it is even more impossible to get the actual values. So not even knowing those values, how can we get the value of the product of the true roots?

Examining the coefficients however, we could get the product or sum of the roots.

For this equation, we can get the value of the product without knowing the values of the true roots.

Suppose that $s$ and $t$ are the two roots for the equation for $u$ above.

Then, $u = s$ or $t$. Thus, we get: $(u - s)(u + t) = 0 \Rightarrow u^2 - (s + t)u + st = 0$.

So we can see that $-(s + t) = \log 6$, and that $st = (\log 2)(\log 3) + a^2$.

Thus, we get: $s + t = -\log 6 = \log 6^{-1}$. What then, is the next?

We have: $u = \log x$, too. So we get: $\log x = s$ or $t$. Thus, we get: $x = 10^s$ or $10^t$.

So we get: $10^s 10^t = 10^{s+t}$ as the product of the two true roots.

We have: $s + t = \log 6^{-1}$, too. So the product is $10^{s+t} = 10^{\log 6^{-1}} = 6^{-1} = \frac{1}{6}$. How come?

Assuming $y = b^{\log_b A}$, and applying $\log_b$ to both sides of the equality, we get:

$$\log_b y = \log_b b^{\log_b A} = (\log_b A)(\log_b b) = \log_b A.$$

Thus, we get: $\log_b y = \log_b A$, so we get: $y = A$.

Now, taking $b = 10$, we get:

$10^{\log 6^{-1}} = 6^{-1}$, since $\log 6^{-1}$ is a common log, and thus, is the same as $\log_{10} 6^{-1}$.

## Suggestions or Solutions
## To the Problem in the Example 2

**Solve the following equation for *x*.**

$$\left(\frac{x}{10^{100}}\right)^u = \frac{\log 10^x}{x} \text{ where } u = (\log x^v - \log x^7 + 10)^3 \text{ where } v = \log x.$$

$$\left(\frac{x}{10^{100}}\right)^u = \frac{\log 10^x}{x} = \frac{x \log 10}{x} = \log 10 = 1 \Rightarrow \left(\frac{x}{10^{100}}\right)^u = 1 \Rightarrow u = 0 \text{ or } \frac{x}{10^{100}} = 1.$$

So first, $\dfrac{x}{10^{100}} = 1 \Rightarrow x = 10^{100}$.

Next, $u = 0 \Rightarrow (\log x^v - \log x^7 + 10)^3 = 0 \Rightarrow \log x^v - \log x^7 + 10 = 0.$

So $v = \log x \Rightarrow \log x^v - \log x^7 + 10 = v \log x - 7 \log x + 10 = (\log x)^2 - 7 \log x + 10 = 0.$

Thus, $(\log x - 5)(\log x - 2) = 0 \Rightarrow \log x = 5 \text{ or } 2 \Rightarrow x = 10^5 \text{ or } 10^2.$

Therefore, $x = 10^{100}, 10^5, \text{ or } 10^2.$

*If not quite sure of the idea behind the processes above, follow the steps below:*

This example looks awfully complicated, doesn't it?

Quite often however, the solution to such a problem is pretty simple.

Examining the equation, we can notice that the right hand side simply reduces to 1.

We can have: $\dfrac{\log 10^x}{x} = \dfrac{x \log 10}{x} = \log 10 = 1.$   Thus, we can see that $\left(\dfrac{x}{10^{100}}\right)^u = 1.$

Thus, we simply get: $u = 0$, or $\dfrac{x}{10^{100}} = 1$.   So we get: $x = 10^{100}$, for now.

Let's next, get this one done: $u = 0$.

We have: $u = (\log x^y - \log x^7 + 10)^3$.

Since $u = 0$, we simply get: $\log x^y - \log x^7 + 10 = 0$.   We have: $v = \log x$, too.   So?

We can get: $\log x^y - \log x^7 + 10 = v \log x - 7 \log x + 10 = (\log x)^2 - 7 \log x + 10 = 0$.

So we get: $(\log x - 5)(\log x - 2) = 0 \Rightarrow \log x = 5$ or $2 \Rightarrow x = 10^5$ or $10^2$.

Therefore, $x = 10^{100}$, $10^5$, or $10^2$.

**In short:**

$\left(\dfrac{x}{10^{100}}\right)^u = \dfrac{\log 10^x}{x} = \dfrac{x \log 10}{x} = \log 10 = 1 \Rightarrow \left(\dfrac{x}{10^{100}}\right)^u = 1 \Rightarrow u = 0$ or $\dfrac{x}{10^{100}} = 1$.

So first, $\dfrac{x}{10^{100}} = 1 \Rightarrow x = 10^{100}$.

Next, $u = 0 \Rightarrow (\log x^y - \log x^7 + 10)^3 = 0 \Rightarrow \log x^y - \log x^7 + 10 = 0$.

So $v = \log x \Rightarrow \log x^y - \log x^7 + 10 = v \log x - 7 \log x + 10 = (\log x)^2 - 7 \log x + 10 = 0$.

Thus, $(\log x - 5)(\log x - 2) = 0 \Rightarrow \log x = 5$ or $2 \Rightarrow x = 10^5$ or $10^2$.

Therefore, $x = 10^{100}$, $10^5$, or $10^2$.

## Examples C on Logarithms

0.  Assuming $a$, $b$, $c$, and $d > 0$, but $\neq 1$, and $b^2 = ac$, show: $\dfrac{\log_a d}{\log_c d} = \dfrac{\log_a d - \log_b d}{\log_b d - \log_c d}$.

1.  Solve $(\log_3 x)\sqrt{\log_x \sqrt{3x}} = -1$.

## Suggestions or Solutions
## To the Problem in the Example 0

Assuming $a$, $b$, $c$, and $d > 0$, but $\neq 1$, and $b^2 = ac$, show: $\dfrac{\log_a d}{\log_c d} = \dfrac{(\log_a d - \log_b d)}{(\log_b d - \log_c d)}$.

On the left hand side, we get: $\dfrac{\log_a d}{\log_c d} = \dfrac{\frac{\log d}{\log a}}{\frac{\log d}{\log c}} = \dfrac{\log c}{\log a} = \log_a c$.

On the right hand side:

$\log_a d - \log_b d = \dfrac{\log d}{\log a} - \dfrac{\log d}{\log b} = \log d \left(\dfrac{1}{\log a} - \dfrac{1}{\log b}\right) = \log d \, \dfrac{\log b - \log a}{(\log a)(\log b)}$.

$\log_b d - \log_c d = \dfrac{\log d}{\log b} - \dfrac{\log d}{\log c} = \log d \left(\dfrac{1}{\log b} - \dfrac{1}{\log c}\right) = \log d \, \dfrac{\log c - \log b}{(\log b)(\log c)}$.

Thus, $\dfrac{\log_a d - \log_b d}{\log_b d - \log_c d} = \dfrac{\log d \, \frac{\log b - \log a}{(\log a)(\log b)}}{\log d \, \frac{\log c - \log b}{(\log b)(\log c)}} = \dfrac{\frac{\log b - \log a}{(\log a)(\log b)}}{\frac{\log c - \log b}{(\log b)(\log c)}} = \dfrac{\log b - \log a}{(\log a)(\log b)} \cdot \dfrac{(\log b)(\log c)}{\log c - \log b}$

$= \dfrac{\log b - \log a}{\log a} \cdot \dfrac{\log c}{\log c - \log b} = \dfrac{\log c}{\log a} \cdot \dfrac{\log b - \log a}{\log c - \log b} = \log_a c \cdot \dfrac{\log b - \log a}{\log c - \log b}$.

So we now, have: $\dfrac{\log_a d}{\log_c d} = \log_a c$, and $\dfrac{\log_a d - \log_b d}{\log_b d - \log_c d} = \log_a c \cdot \dfrac{\log b - \log a}{\log c - \log b}$.

Thus, we need to have: $\dfrac{\log b - \log a}{\log c - \log b} = 1$. And we have: $b^2 = ac$. So we get:

$\log b^2 = \log ac \Rightarrow 2 \log b = \log ac = \log a + \log c \Rightarrow 2 \log b = \log a + \log c$

$\Rightarrow \log b - \log a = \log c - \log b$.

Thus, we get: $\dfrac{\log b - \log a}{\log c - \log b} = 1$.

So we now, have $\dfrac{\log_a d}{\log_c d} = \log_a c$, and also, $\dfrac{\log_a d - \log_b d}{\log_b d - \log_c d} = \log_a c$.

Therefore, $\dfrac{\log_a d}{\log_c d} = \dfrac{(\log_a d - \log_b d)}{(\log_b d - \log_c d)}$ where $b^2 = ac$.

*If not quite sure of the idea behind the processes above, follow the steps below:*

A log can be put in a fractional form.

For instance, we can modify $\log_x y$ to $\dfrac{\log y}{\log x}$.   That is, we have: $\log_x y = \dfrac{\log y}{\log x}$.

And we call the equality a log identity. So?

So we can use such an identity, together with other identities on logs, of course, when working with a log expression like those in the equality above.

Examining the equality given, we can see that all the logs have the same antilog, which is **d**. Thus, we can expect that **d** can be removed (actually, canceled out). It is like canceling out processes in fractional multiplications.

For instance, we can have: $\dfrac{3}{4} \cdot \dfrac{4}{5} = \dfrac{3}{5}$, where 4s get canceled out.

So how are we going to get this problem done?

If we can modify both sides of the equality so that both sides get turned out to be the same expression, the equality is proven. And doing the modifications, we are going to use the identity above. So let's now, begin with the left hand side.

$$\frac{\log_a d}{\log_c d} = \frac{\frac{\log d}{\log a}}{\frac{\log d}{\log c}} = \frac{\log d}{\log a} \cdot \frac{\log c}{\log d} = \frac{\log c}{\log a} = \log_a c.$$

Next, let's move on to the right hand side, and begin with the numerator.

$$\log_a d - \log_b d = \frac{\log d}{\log a} - \frac{\log d}{\log b} = \log d \left( \frac{1}{\log a} - \frac{1}{\log b} \right) = \log d \, \frac{\log b - \log a}{(\log a)(\log b)}.$$

And next, moving on to the denominator, we can get:

$$\log_b d - \log_c d = \frac{\log d}{\log b} - \frac{\log d}{\log c} = \log d \left( \frac{1}{\log b} - \frac{1}{\log c} \right) = \log d \, \frac{\log c - \log b}{(\log b)(\log c)}.$$

Thus, we get:

$$\frac{\log_a d - \log_b d}{\log_b d - \log_c d} = \frac{\log d \, \frac{\log b - \log a}{(\log a)(\log b)}}{\log d \, \frac{\log c - \log b}{(\log b)(\log c)}} = \frac{\frac{\log b - \log a}{(\log a)(\log b)}}{\frac{\log c - \log b}{(\log b)(\log c)}} = \frac{\log b - \log a}{(\log a)(\log b)} \cdot \frac{(\log b)(\log c)}{\log c - \log b}$$

$$= \frac{\log b - \log a}{\log a} \cdot \frac{\log c}{\log c - \log b} = \frac{\log c}{\log a} \cdot \frac{\log b - \log a}{\log c - \log b} = \log_a c \cdot \frac{\log b - \log a}{\log c - \log b}.$$

So we now, have: $\dfrac{\log_a d}{\log_c d} = \log_a c$, and $\dfrac{\log_a d - \log_b d}{\log_b d - \log_c d} = \log_a c \cdot \dfrac{\log b - \log a}{\log c - \log b}$.

Thus, we get: $\dfrac{\log_a d}{\log_c d} = \dfrac{(\log_a d - \log_b d)}{(\log_b d - \log_c d)} \Rightarrow \log_a c = \log_a c \cdot \dfrac{\log b - \log a}{\log c - \log b}$.  So what?

We want to show that $\dfrac{\log b - \log a}{\log c - \log b} = 1$.  How though?

We can show this, instead: $\log b - \log a = \log c - \log b$.

We can't just show that of course. We need to have something to work with.

And we haven't used yet something that is given. What then, is it?

It is the supposition that $b^2 = ac$, and we want to take advantage of it. How can we use it though?

We are now working with logs.    So?

So taking advantage of the supposition, we can apply logs to it.    Then, we get:

$\log b^2 = \log ac = \log a + \log c \Rightarrow 2 \log b = \log a + \log c$.

Thus, we get: $\log b = \log a - \log c - \log b \Rightarrow \log b - \log a = \log c - \log b$.

**In short:**

On the left hand side, we get: $\dfrac{\log_a d}{\log_c d} = \dfrac{\frac{\log d}{\log a}}{\frac{\log d}{\log c}} = \dfrac{\log c}{\log a} = \log_a c$.

On the right hand side:

$\log_a d - \log_b d = \dfrac{\log d}{\log a} - \dfrac{\log d}{\log b} = \log d(\dfrac{1}{\log a} - \dfrac{1}{\log b}) = \log d \dfrac{\log b - \log a}{(\log a)(\log b)}$.

$\log_b d - \log_c d = \dfrac{\log d}{\log b} - \dfrac{\log d}{\log c} = \log d(\dfrac{1}{\log b} - \dfrac{1}{\log c}) = \log d \dfrac{\log c - \log b}{(\log b)(\log c)}$.

Thus, $\dfrac{\log_a d - \log_b d}{\log_b d - \log_c d} = \dfrac{\log d \, \frac{\log b - \log a}{(\log a)(\log b)}}{\log d \, \frac{\log c - \log b}{(\log b)(\log c)}} = \dfrac{\frac{\log b - \log a}{(\log a)(\log b)}}{\frac{\log c - \log b}{(\log b)(\log c)}} = \dfrac{\log b - \log a}{(\log a)(\log b)} \cdot \dfrac{(\log b)(\log c)}{\log c - \log b}$

$= \dfrac{\log b - \log a}{\log a} \cdot \dfrac{\log c}{\log c - \log b} = \dfrac{\log c}{\log a} \cdot \dfrac{\log b - \log a}{\log c - \log b} = \log_a c \cdot \dfrac{\log b - \log a}{\log c - \log b}.$

So we now, have: $\dfrac{\log_a d}{\log_c d} = \log_a c$, and $\dfrac{\log_a d - \log_b d}{\log_b d - \log_c d} = \log_a c \cdot \dfrac{\log b - \log a}{\log c - \log b}.$

Thus, we need to have: $\dfrac{\log b - \log a}{\log c - \log b} = 1.$ And we have: $b^2 = ac.$

So we get: $\log b^2 = \log ac \Rightarrow 2 \log b = \log ac = \log a + \log c \Rightarrow 2 \log b = \log a + \log c$

$\Rightarrow \log b - \log a = \log c - \log b.$

Thus, we get: $\dfrac{\log b - \log a}{\log c - \log b} = 1.$

So we now, have $\dfrac{\log_a d}{\log_c d} = \log_a c$, and also, $\dfrac{\log_a d - \log_b d}{\log_b d - \log_c d} = \log_a c.$

Therefore, $\dfrac{\log_a d}{\log_c d} = \dfrac{(\log_a d - \log_b d)}{(\log_b d - \log_c d)}$ where $b^2 = ac.$

## Suggestions or Solutions
## To the Problem in the Example 1

**Solve for $x$ in the equation, $(\log_3 x)\sqrt{\log_x \sqrt{3x}} = -1$.**

$$(\log_3 x)\sqrt{\log_x \sqrt{3x}} = -1 \Rightarrow \sqrt{\log_x \sqrt{3x}} = -\log_x 3 \Rightarrow \log_x \sqrt{3x} = (\log_x 3)^2$$

$$\Rightarrow \tfrac{1}{2}\log_x 3x = (\log_x 3)^2 \Rightarrow \log_x 3x = 2(\log_x 3)^2 \Rightarrow \log_x 3 + 1 = 2(\log_x 3)^2.$$

Setting $\log_x 3 = t$, we get: $2t^2 - t - 1 = 0 \Rightarrow (2t+1)(t-1) = 0 \Rightarrow t = -\tfrac{1}{2}$ or $1$.

$t = -\tfrac{1}{2} \Rightarrow \log_x 3 = -\tfrac{1}{2} \Rightarrow 3 = x^{-\frac{1}{2}} \Rightarrow x^{-1} = 9 \Rightarrow x = 9^{-1}$.

$t = 1 \Rightarrow \log_x 3 = 1 \Rightarrow x = 3$.

$x = 3 \Rightarrow (\log_3 x)\sqrt{\log_x \sqrt{3x}} = (\log_3 3)\sqrt{\log_3 \sqrt{9}} = \sqrt{\log_3 \sqrt{9}} = \sqrt{\log_3 3} = 1 \neq -1$.

Thus, $x = 3$ is not allowed.

$x = 9^{-1} \Rightarrow (\log_3 x)\sqrt{\log_x \sqrt{3x}} = (\log_3 9^{-1})\sqrt{\log_{9^{-1}} \sqrt{3 \cdot 9^{-1}}} = -2\sqrt{\log_{9^{-1}} \sqrt{3^{-1}}}$

$= -2\sqrt{\log_{3^{-2}} 3^{-\frac{1}{2}}} = -2\sqrt{\frac{-\frac{1}{2}}{-2}} = -2\sqrt{\frac{1}{4}} = -2 \cdot \tfrac{1}{2} = -1$.  Therefore, $x = 9^{-1}$.

*If not quite sure of the idea behind the processes above, follow the steps below:*

It looks like we have three unknowns in one equation.
They are the same though, of course, and $x$ is the unknown.
In the equation, the same unknown $x$ is in three different places.

It is inside a log, that is, it's an antilog, and is inside a square root sign, and also, it is used as a base in a log, too.   So how are we going to do with it?

Basically, solving an equation, we are isolating the unknown so that we can eventually get a form as follows: $x =$ a number. How then, can we get it isolated?

We can remove the square root sign. How though?

First, we can isolate the square root part by moving $\mathbf{log_3}\, x$ to the right hand side. How?

We can divide both sides by $\mathbf{log_3}\, x$ respectively.
So doing the divisions, and doing some more algebra, we can get:

$$\sqrt{\log_x \sqrt{3x}} = \frac{-1}{\log_3 x} = \frac{-1}{\frac{\log x}{\log 3}} = -\frac{\log 3}{\log x} = -\log_x 3 \Rightarrow \sqrt{\log_x \sqrt{3x}} = -\log_x 3.$$

Next, squaring both sides, we can remove the outer square root sign. Then, we get:

$$\log_x \sqrt{3x} = (\log_x 3)^2.$$

Then, negative cases can get involved due to squaring, so we don't want to forget to check to see if all the solutions are appropriate. Now, moving on to the next step, we get:

$$\log_x \sqrt{3x} = \tfrac{1}{2}\log_x 3x = (\log_x 3)^2 \Rightarrow \log_x 3x = 2(\log_x 3)^2. \quad \text{What then?}$$

We know: $\mathbf{log_x\, 3x = log_x\, 3 + log_x\, x = log_x\, 3 + 1}$. So we get: $\mathbf{log_x\, 3 + 1 = 2(log_x\, 3)^2}$.

What equation then, do we get? Isn't it getting more complicated, is it?

We just get a quadratic equation, which is usually easy to solve. What quadratic equation though?

It is a quadratic equation for $\log_x 3$, and not just $x$.    How then, can we get it solved?

We can set, for instance, $u = \log_x 3$.
Then, we can get a quadratic equation for $u$.
So we solve that quadratic equation first.    And then, we can solve $u = \log_x 3$ for $x$.

So setting now, $u = \log_x 3$, and solving the equation, we get:

$$\log_x 3 + 1 = 2(\log_x 3)^2 \Rightarrow u + 1 = 2u^2 \Rightarrow 2u^2 - u - 1 = 0 \Rightarrow (2u+1)(u-1) = 0$$

$$\Rightarrow u = -\tfrac{1}{2} \text{ or } 1. \quad \text{Then, we get two equations.}$$

One is: $-\tfrac{1}{2} = \log_x 3$.  And the other is: $1 = \log_x 3$.

So first, by the definition for logs, we get: $\log_x 3 = -\tfrac{1}{2} \Rightarrow 3 = x^{-\frac{1}{2}} \Rightarrow x^{-1} = 9 \Rightarrow x = 9^{-1}$.

And next, by the definition again, we get: $\log_x 3 = 1 \Rightarrow x = 3$.

That's not it though. The quadratic equation we just have solved is not really the original equation given. So we want to see if the two roots are OK.

So assuming first, $x = 3$, we get:

$$x = 3 \Rightarrow (\log_3 x)\sqrt{\log_x \sqrt{3x}} = (\log_3 3)\sqrt{\log_3 \sqrt{9}} = \sqrt{\log_3 \sqrt{9}} = \sqrt{\log_3 3} = 1 \neq -1.$$

Thus, $x = 3$ is not allowed.

And next, $x = 9^{-1} \Rightarrow (\log_3 x)\sqrt{\log_x \sqrt{3x}} = (\log_3 9^{-1})\sqrt{\log_{9^{-1}} \sqrt{3 \cdot 9^{-1}}} = -2\sqrt{\log_{9^{-1}} \sqrt{3^{-1}}}$

$$= -2\sqrt{\log_{3^{-2}} 3^{-\frac{1}{2}}} = -2\sqrt{\tfrac{-\frac{1}{2}}{-2}} = -2\sqrt{\tfrac{1}{4}} = -2 \cdot \tfrac{1}{2} = -1.$$

Therefore, $x = 9^{-1}$ is the solution.

**In short:**

$$(\log_3 x)\sqrt{\log_x \sqrt{3x}} = -1 \Rightarrow \sqrt{\log_x \sqrt{3x}} = -\log_x 3 \Rightarrow \log_x \sqrt{3x} = (\log_x 3)^2$$

$$\Rightarrow \tfrac{1}{2}\log_x 3x = (\log_x 3)^2 \Rightarrow \log_x 3x = 2(\log_x 3)^2 \Rightarrow \log_x 3 + 1 = 2(\log_x 3)^2.$$

Setting now, $\log_x 3 = t$, we can get: $2t^2 - t - 1 = 0 \Rightarrow (2t + 1)(t - 1) = 0 \Rightarrow t = -\tfrac{1}{2}$ or $1$.

$$t = -\tfrac{1}{2} \Rightarrow \log_x 3 = -\tfrac{1}{2} \Rightarrow 3 = x^{-\frac{1}{2}} \Rightarrow x^{-1} = 9 \Rightarrow x = 9^{-1}.$$

$$t = 1 \Rightarrow \log_x 3 = 1 \Rightarrow x = 3.$$

Now, we can check to see if the values are OK using this: $(\log_3 x)\sqrt{\log_x \sqrt{3x}} = -1$.

$$x = 3 \Rightarrow (\log_3 x)\sqrt{\log_x \sqrt{3x}} = (\log_3 3)\sqrt{\log_3 \sqrt{9}} = \sqrt{\log_3 \sqrt{9}} = \sqrt{\log_3 3} = 1 \neq -1.$$

Thus, $x = 3$ is not allowed.

$$x = 9^{-1} \Rightarrow (\log_3 x)\sqrt{\log_x \sqrt{3x}} = (\log_3 9^{-1})\sqrt{\log_{9^{-1}} \sqrt{3 \cdot 9^{-1}}} = -2\sqrt{\log_{9^{-1}} \sqrt{3^{-1}}}$$

$$= -2\sqrt{\log_{3^{-2}} 3^{-\frac{1}{2}}} = -2\sqrt{\frac{-\frac{1}{2}}{-2}} = -2\sqrt{\frac{1}{4}} = -2 \cdot \tfrac{1}{2} = -1.$$

Therefore, $x = 9^{-1}$ is the solution.

## Examples D on Logarithms

0.   Using **log 4.6065 = 0.6634**, and **log 1.908 = 0.2805**, evaluate $\textbf{1.908}^{\textbf{-1.2}}$.

1.   Assuming $u$ is the index (or characteristic) of **log** $x$, and $v$ is that of $\textbf{log}\frac{10}{x}$, find $x$ that maximizes $u^2 - 2v^2$, and find the maximum value.

## Suggestions or Solutions
## To the Problem in the Example 0

**Using log 4.6065 = 0.6634 and log 1.908 = 0.2805, evaluate $1.908^{-1.2}$.**

**log $1.908^{-1.2}$ = -1.2 log 1.908**. And we have: **log 1.908 = 0.2805**.

So we get: **-1.2 log 1.908 = -1.2(0.2805) = -0.3366 = -1 + 1 - 0.3366 = -1 + 0.6634**.

And we have: **log 4.6065 = 0.6634**.

So the value of **$1.908^{-1.2}$** is **0.46065**.

*If not quite* sure *of the idea behind the processes above, follow the steps below:*

In this problem, we are given a means to work with. What then, is the means?

It is a set of two logs and their values.

One is: **log 4.6065 = 0.6634**, and the other is: **log 1.908 = 0.2805**.

So what do we do with those?

Since we need to work with log-values, we may want to take a log of the value to be evaluated, and the log is a common log, of course. And then, we apply the means to it.

So to begin with, taking the common of **$1.908^{-1.2}$**, we get: **log $1.908^{-1.2}$ = -1.2 log 1.908**.

And we have: **log 1.908 = 0.2805**.

So we get: **-1.2 log 1.908 = -1.2(0.2805) = -0.3366**.  What do we do with it then?

The log above is a common log, so it can be put in terms of the characteristic and mantissa. And we know: $0 \leq \textbf{mantissa} < 1$.

So we can get: $\log 1.908^{-1.2} = -0.3366 = -1 + 1 - 0.3366 = -1 + 0.6634$.

Now, we have another piece of the means, which is: $\log 4.6065 = 0.6634$, which is the same as the mantissa above.    So what are we going to do with it?

Suppose now, that $\log A = c + m$, where $A > 0$, $c$ is an integer called the characteristic, and $m$ is the mantissa.

Then first, the highest place value in $A$ is $10^c$.
In particular, if $c < 0$, the first nonzero appears in the $c^{th}$ digit below the decimal point.

Next, we have another fact as follows:

  • If we get the same mantissas after taking common logs of two numbers, the two numbers have the same sequence of digits.

So what are we going to do with the fact?

We have two common logs, which are $\log 1.908^{-1.2}$ and $\log 4.6065$, both of which share the same mantissa, which is **0.6634**.    So?

So $1.908^{-1.2}$ has the same sequence of digits as **4.6065** has. Thus, the sequence of digits in the number $1.908^{-1.2}$ is $\{4, 6, 0, 6, 5\}$.

Next, the characteristic of $\log 1.908^{-1.2}$ is -1, so the first nonzero appears in the first digit below the decimal point.

Therefore, the value of $1.908^{-1.2}$ is **0.46065**.

**In short:**

**log 1.908$^{-1.2}$ = -1.2 log 1.908**.   And we have: **log 1.908 = 0.2805**.

So we get: **-1.2 log 1.908 = -1.2(0.2805) = -0.3366 = -1 + 1 - 0.3366 = -1 + 0.6634**.

And we have: **log 4.6065 = 0.6634**.

So the value of **1.908$^{-1.2}$** is **0.46065**.

## Suggestions or Solutions
## To the Problem in the Example 1

**Assuming $u$ is the index (or characteristic) of $\log x$, and $v$ is that of $\log \frac{10}{x}$, find $x$ that maximizes $u^2 - 2v^2$, and find the maximum value.**

Suppose $\log x = u + m$, where $0 \le m < 1$.

Then, we get: $\log \frac{10}{x} = \log 10 - \log x = 1 - u - m \Rightarrow \log \frac{10}{x} = 1 - u - m$.

If $m = 0$, we get: $\log \frac{10}{x} = 1 - u - m = 1 - u + 0$.   So the mantissa in this case is 0.

If $0 < m < 1$, we get: $\log \frac{10}{x} = 1 - u - m = 1 - u - 1 + 1 - m = -u + 1 - m$.   So in this case the mantissa is: $1 - m$.

And we know $v$ is the index of $\log \frac{10}{x}$.

So beginning with the case where $m = 0$, and $v = 1 - u$, we get:

$u^2 - 2v^2 = u^2 - 2(1 - u)^2 = u^2 - 2(1 - 2u + u^2) = -u^2 + 4u - 2 = -(u^2 - 4u + 2)$
$= -(u^2 - 4u + 4 - 4 + 2) = -(u^2 - 4u + 4 - 2) = -(u - 2)^2 + 2 \le 2$.

So if $m = 0$, when $u = 2$, $(u^2 - 2v^2)$ gets 2 as its maximum.

And we know: $\log x = u + m$, where $0 \le m < 1$.

So when $u = 2$, we get: $\log x = 2 + 0 = 2$, and thus, we get: $x = 100$.

And next, moving on to the case where $0 < m < 1$, $v = -u$, we get:

$u^2 - 2v^2 = u^2 - 2u^2 = -u^2 \le 0$.

So if $0 < m < 1$, when $u = 0$, $(u^2 - 2v^2)$ gets 0 as its maximum.

So when $u = 0$, we get: $\log x = 0 + m = m$, and thus, we get: $x = 10^m$ where $0 < m < 1$.

So when $u = 0$, we get: $1 < x < 10$.

Putting thus, threads together, we can say that when $x = 100$, $u^2 - 2v^2$ gets its maximum, which is 2.

*If not quite sure of the idea behind the processes above, follow the steps below:*

To begin with, we can put $\log_b A$ the way below:

$\log_b A = n + m$, where $n$ is an integer called the index, $0 \leq m < 1$, and $m$ is called the mantissa.

And in particular, if the base $b = 10$, the log is called a common log, and $n$ is often called the characteristic. Then, the highest place value in $A$ is $10^n$.

And we know the fact that $u$ is the characteristic of $\log x$.

So we can put $\log x$ this way: $\log x = u + m$, where $0 \leq m < 1$. What then, about $\log \frac{10}{x}$?

We can get: $\log \frac{10}{x} = \log 10 - \log x = 1 - u - m \Rightarrow \log \frac{10}{x} = 1 - u - m$.

Note that $m$ is the mantissa of $\log x$, and not that of $\log \frac{10}{x}$. So?

So we need to determine the mantissa of $\log \frac{10}{x}$. And we have two cases to consider.

One is: $m = 0$, and the other is: $0 < m < 1$.

If $m = 0$, we get: $\log \frac{10}{x} = 1 - u - m = 1 - u + 0$.

So the mantissa of $\log \frac{10}{x}$ is simply 0. And of course, the index is: $1 - u$.

If however, $0 < m < 1$, we get: $\log \frac{10}{x} = 1 - u - m = 1 - u - 1 + 1 - m$.

So we get: $\log \frac{10}{x} = -u + (1 - m)$, and $0 < 1 - m < 1$.

So the mantissa of $\log \frac{10}{x}$ is $(1 - m)$. And of course, the index is: $-u$.

The problem says though, the index of $\log \frac{10}{x}$ is $v$. So?

So we have two cases for $v$:

- If $m = 0$, the index of $\log \frac{10}{x}$ is $1 - u$, so in this case, we have: $v = 1 - u$.

- And next, if $0 < m < 1$, the index of $\log \frac{10}{x}$ is $-u$, so in this case, we have: $v = -u$.

So in sum, we have two cases to consider doing this problem.

One is the case: $m = 0$, and $v = 1 - u$.    And the other is the case: $0 < m < 1$ and $v = -u$.

Let's now find $x$ that maximizes $u^2 - 2v^2$, and find the maximum value.

Beginning with the case where $m = 0$, and $v = 1 - u$, we get:

$u^2 - 2v^2 = u^2 - 2(1 - u)^2 = u^2 - 2(1 - 2u + u^2) = -u^2 + 4u - 2 = -(u^2 - 4u + 2)$
$= -(u^2 - 4u + 4 - 4 + 2) = -(u^2 - 4u + 4 - 2) = -(u - 2)^2 + 2 \leq 2$.

So in this case, 2 is the maximum of $u^2 - 2v^2$.
And when it gets its maximum, we get: $u = 2$.

And we know: $\log x = u + m$, where $0 \leq m < 1$.

So at the maximum, we get: $\log x = u + m = 2 + 0 = 2$, and therefore, $x = 100$.


And next, moving on to the case where $0 < m < 1$ and $v = -u$, we get:

$u^2 - 2v^2 = u^2 - 2u^2 = -u^2 \leq 0$.

Thus, in this case, when $u = 0$, $(u^2 - 2v^2)$ gets 0 as its maximum.

Then, when $u = 0$, we get: $\log x = u + m = m$.

So by the definition for logs, we get: $x = 10^m$, where $0 < m < 1$, and thus, $1 < x < 10$.

Thus, putting threads together, when $x = 100$, $u^2 - 2v^2$ gets 2, which is the maximum.

**In short:**

Suppose $\log x = u + m$, where $0 \leq m < 1$.

Then, we get: $\log \frac{10}{x} = \log 10 - \log x = 1 - u - m \Rightarrow \log \frac{10}{x} = 1 - u - m$.

If $m = 0$, we get: $\log \frac{10}{x} = 1 - u - m = 1 - u + 0$.    So the mantissa in this case is 0.

If $0 < m < 1$, we get: $\log \frac{10}{x} = 1 - u - m = 1 - u - 1 + 1 - m = -u + 1 - m$.    So in this case the mantissa is: $1 - m$.

And we know $v$ is the index of $\log \frac{10}{x}$.

So beginning with the case where $m = 0$, and $v = 1 - u$, we get:

$u^2 - 2v^2 = u^2 - 2(1 - u)^2 = u^2 - 2(1 - 2u + u^2) = -u^2 + 4u - 2 = -(u^2 - 4u + 2)$

$= -(u^2 - 4u + 4 - 4 + 2) = -(u^2 - 4u + 4 - 2) = -(u - 2)^2 + 2 \leq 2$.

So if $m = 0$, when $u = 2$, $(u^2 - 2v^2)$ gets 2 as its maximum.
And we know: $\log x = u + m$, where $0 \leq m < 1$.
So when $u = 2$, we get: $\log x = 2 + 0 = 2$, and thus, we get: $x = 100$.

And next, moving on to the case where $0 < m < 1$, $v = -u$, we get:

$u^2 - 2v^2 = u^2 - 2u^2 = -u^2 \leq 0$.

So if $0 < m < 1$, when $u = 0$, $(u^2 - 2v^2)$ gets 0 as its maximum.
And we know: $\log x = u + m$, where $0 \leq m < 1$.
So when $u = 0$, we get: $\log x = 0 + m = m$, and thus, we get: $x = 10^m$ where $0 < m < 1$.
So when $u = 0$, we get: $1 < x < 10$.

Putting thus, threads together, we can say that when $x = 100$, $u^2 - 2v^2$ gets its maximum, which is 2.

# c. **Approximations**

To begin with, what is a log-value?

Taking the value of the log of a number to a base, we get the log-value to the base.
So a log-value depends on the base to which we take a log of the number.
For instance, the log of 16 to base 2 is 4, but the log of 16 to base 4 is 2, so in this case, 4 is the log-value to base 2, and 2 is the log-value to base 4. So what is a log-value?

It's an exponent that we use expressing a number in a power of a particular base.
Quite usually though, saying just a log, we mean a common log, where the base is 10.
So saying just a log-value, we usually mean the value of a common log.
How do we get such a value, though?

We can get it using a calculator, of course. What if no calculator or none of that sort?

If we want to get a log-value as the value of **log 2**, a log table can help.
Usually, such a log table is for values of common logs.

So if we want to get the value of the common log of a particular number, we can use a log table, which is a list of some antilogs and their log-values, that is, some numbers and the values of the common logs of those numbers. And thus, if the list includes the particular number we want to take the common log of, we can get the log-value at once.

Usually though, log-values listed in a log table are only mantissas, and antilogs are only between 1 and 10. (If not sure of mantissas, refer to the section, **Common Logs**.)

Thus, getting the value of a log of a particular number using a log table, we can get a problem where the particular number is not listed in the table. What then, can we do?

We can an *approximation*. We can get a value close enough to the value of the log of the particular number. So let's see now, how we can get the approximation.

Suppose now, *P* is positive, the table has two antilogs *A* and *B*, which are sufficiently close to *P*, and $A < P < B$, and we want to get the value of **log P**, but *P* is not in the table.

Then, we can get an approximate value of **log P**, and can get it the way below.

First, we can get the values of **log A** and **log B**, since they are listed in the table.

Next, putting in a graph the two log-values, along with a part of a log-curve, we can get:

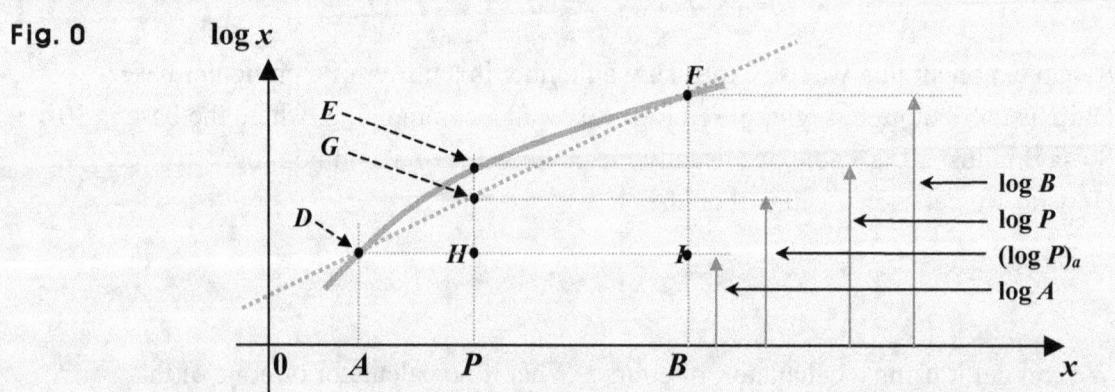

**Fig. 0**

The distance from *E* to *G* should be sufficiently small, which means negligibly small. Then, we can get a good approximation by means of ratios that apply to similar triangles.

To begin with, we can see that two triangles **DIF** and **DHG** are similar to each other.

So next, assuming **AB** is the distance from *A* to *B*, we can see that:
**AB : AP = DI : DH = FI : GH**.

That is to say that we get: $\dfrac{AB}{AP} = \dfrac{DI}{DH} = \dfrac{FI}{GH}$.   So we get: $\dfrac{AB}{AP} = \dfrac{FI}{GH}$.

And assuming $(\log P)_a$ is an approximation of $\log P$, we can get: $GH + \log A = (\log P)_a$.

So finding $GH$, we can an approximate value of $\log P$.   How?

We can get it from the ratio equation above.

So first, we can get: $\dfrac{AB}{AP} = \dfrac{FI}{GH} \Rightarrow GH \cdot \dfrac{AB}{AP} = FI \Rightarrow GH = FI \cdot \dfrac{AP}{AB}$.

And next, we can get: $FI = \log B - \log A$, $AP = P - A$, and $AB = B - A$.

So we get: $GH = (\log B - \log A) \cdot \dfrac{P - A}{B - A}$.

Now, conversely, what if a log-value is given, and we want to get the antilog, but the log-value is not in the table?

For instance, using a log table, we want to find $x$ for which we get: $0.3581 = \log x$, but the table does not have $0.3581$.

The exact solution is $x = 10^{0.3581}$, and using a calculator, we can get the value of $x$, that is, the value of the power $10^{0.3581}$.   What if however, no calculator is allowed?

We can get an approximate value, called an approx value, for short.
And we can get it using a log-table, too.   How?

We can use basically the same idea as the one for an approximation of a log-value.

So suppose now, $p = \log A$, the table has two log-values $m$ and $n$, which are sufficiently close to $p$, where $m < p < n$, and we want to get $A$, but $A$ is not in the table.

Then, we can get an *approx value* of **A**, and can get it the way below.

To begin with, we are given two antilogs, which are the values of $10^m$ and $10^n$, for they are listed in the table.

Next, setting: $M = 10^m$, and $N = 10^n$, and putting in a graph the two antilogs **M** and **N**, along with a part of a log-curve, we get:

**Fig. 1.**     The distance between **H** and **I** should be small enough.

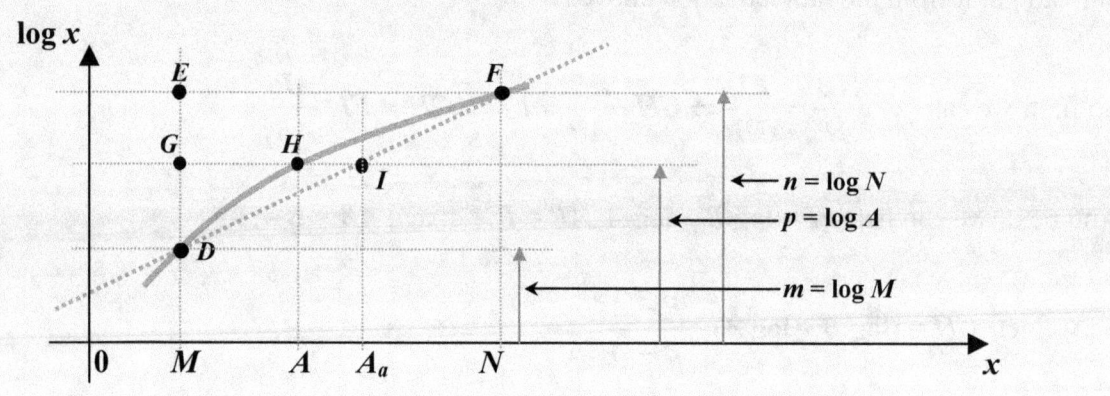

Then again, we can get the approximation of **A** using ratios apply to similar triangles.

First, we can see that two triangles **DGI** and **DEF** are similar to each other.

So next, we can see that: $DG : DE = GI : EF$. That is, we can get: $\dfrac{DG}{DE} = \dfrac{GI}{EF}$.

And assuming $A_a$ is an approximation of **A**, we can get: $M + GI = A_a$.
So finding **GI** using the ratio equation above, we can an approximation of **A**.

Thus, to begin with, we can get: $\dfrac{DG}{DE} = \dfrac{GI}{EF} \Rightarrow GI = EF \cdot \dfrac{DG}{DE}$.

And next, we can get: $EF = N - M$, $DG = \log A - \log M$, and $DE = \log N - \log M$.

So we get: $GI = (N - M) \cdot \dfrac{\log A - \log M}{\log N - \log M}$.

Now, usually, log-values listed in a log table are mantissas, which are between 0 and 1, and antilogs are numbers between 1 and 10.

Then, what if the log-value we want to get an approximation of is bigger than 1 or the antilog we need to get an approximation of is larger than 10 or between 0 and 1?

For instance, the log-value can be between 23.4 and 23.5, and the antilog can be between 87 and 88, or between 0.123 and 0.124.    What then, can we do?

Shifting the entire digits in a number, and then, taking the log of the new number, we get a new log-value, but the new log-value still has the same mantissa.    Not quite sure?

We know a log-value is composed of an integer and a fractional number.
The integer is called the characteristic, and the fractional number is between 0 and 1, can be 0, too, and is called the mantissa.
(If not sure of characteristics, and mantissas, refer to the section, **Common Logs**.)

So a log-value is made of a characteristic and a mantissa, and we can put a log of $A$ in such a way as follows:

$\log A = c + m$, where $c$ is the characteristic, $m$ is the mantissa, and $0 \leq m < 1$.   So what?

Suppose now, if we take a log of a number, the characteristic is 0, but the mantissa is not.
Suppose next, we shift the entire digits in the number.
That is, we move the decimal point several digits to the left or right.
In other words, $A$ gets multiplied or divided by a power of 10.

Then, we get a new number, of course, but the new number still has the same sequence of digits. That is, the new number keeps the same sequence of number-alphabets.
What number-alphabets?

Number-alphabets are 1-digit numbers as 0, 3, and 8. For details on number-alphabets, refer to **BASES** in **NUMBER SYSETMS**.

So if we take the log of the new number, we get a new characteristic, but the mantissa is the same. That is to say that, *c* changes, but *m* remains the same.

(Why the same mantissa? Refer to the section, **Common Logs**.)

Therefore, using such a shifting method, together with the approximation method, we can get the approximations of the numbers that are larger or smaller than the ones listed in a log table.

For instance, we can have: **log 2.34 = 0.3692** in a log table.

And shifting the entire digits in 2.34 to the left by 3 decimal places, we get 2340.

That is, moving the decimal point 3 digits to the right, we get 2340.

In other words, multiplying 2.34 by $10^3$, we get 2340.

Then, the highest place value in 2340 is $10^3$, where the exponent 3 is the characteristic.

So we get: **log 2340 = 3.3692**.

Shifting, for another instance, the entire digits in 23.4 to the right by 2 decimal places, we get 0.234.

That is, moving the decimal point 2 digits to the left, we get 0.234.

In other words, dividing 23.4 by $10^2$, we get 0.234.

So the highest place value in 0.234 is $10^{-1}$, where the exponent -1 is the characteristic.

Thus, we get: **log 0.234 = $\bar{1}$.3692**, which means however, not −1.3692 but −1 + 0.3692.

So if using a calculator, we get: **log 0.234 = -0.6308**, which is equivalent to $\bar{1}$.3692.

(Why the bar on top of 1 though? Refer to the section, **Common Logs**.)

Now, let's do some examples.

## Examples 1 in Approximations

0.   Using **log 5.72 = 0.7574** and **log 5.73 = 0.7582**, find an approximation of **log 5.724** without calculator.

1.   Using **log 5.72 = 0.7574** and **log 5.73 = 0.7582**, find an approximation of the antilog $A$ in **log $A$ = 0.7579** with no calculator.

## Suggestions or Solutions
## To the Problem in the Example 0

**Using log 5.72 = 0.7574 and log 5.73 = 0.7582, find an approximation of log 5.724 without a calculator.**

Setting $A = 5.72$, and $B = 5.73$, and putting all the values in a graph, we can get:

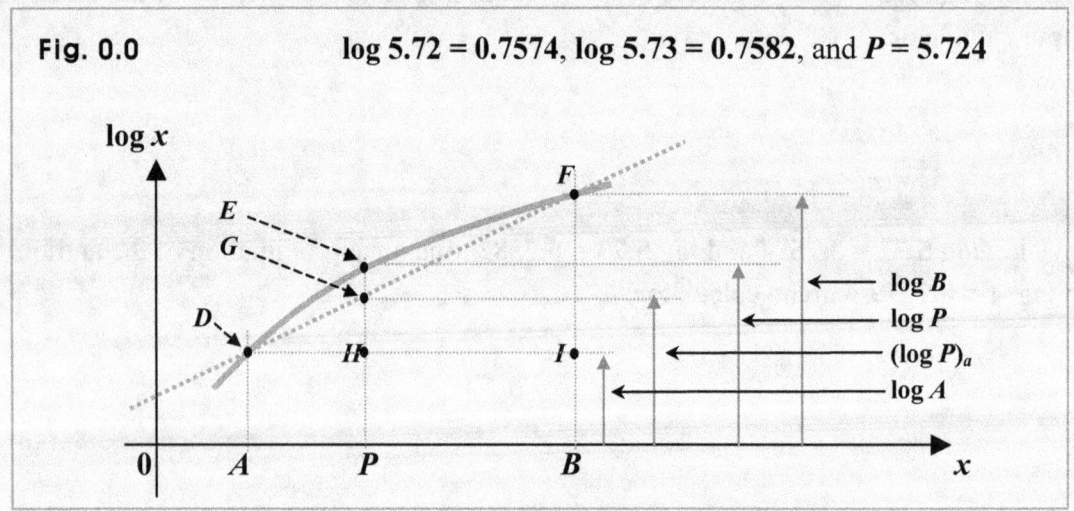

**Fig. 0.0**          **log 5.72 = 0.7574, log 5.73 = 0.7582, and $P = 5.724$**

Since **FI** is parallel to **GH**, two triangles **DIF** and **DHG** are similar to each other.

So setting: **GH = d**, and letting **(log P)$_a$** be the approximation, we can set:

**(log P)$_a$ = log A + d**, where $d = \frac{P-A}{B-A} \cdot (\log B - \log A)$.

We have: $P - A = 5.724 - 5.72 = 0.004$, $B - A = 5.73 - 5.72 = 0.01$, and

**log B − log A = 0.7582 − 0.7574 = 0.0008**.

So we get: $d = \frac{P-A}{B-A} \cdot (\log B - \log A) = \frac{0.004}{0.01} \cdot 0.0008 = 0.4 \cdot 0.0008 = 0.00032$.

Thus, **log 5.724 $\cong$ (log P)$_a$ = log A + d = 0.7574 + 0.00032 = 0.75772 $\cong$ 0.7577**.

*If not quite sure of the idea behind the processes above, follow the steps below:*

Let's first, set: $A = 5.72$, $B = 5.73$, and $P = 5.724$.

Next, taking the log of both sides in each equality, we get:

$\log A = 0.7574$, and $\log B = 0.7582$. That is, $\log 5.72 = 0.7574$, and $\log 5.73 = 0.7582$

And putting next, the values above in a graph, we can get a graph as shown below.

**Fig. 0.1**                 $\log 5.72 = 0.7574$, $\log 5.73 = 0.7582$, and $P = 5.724$

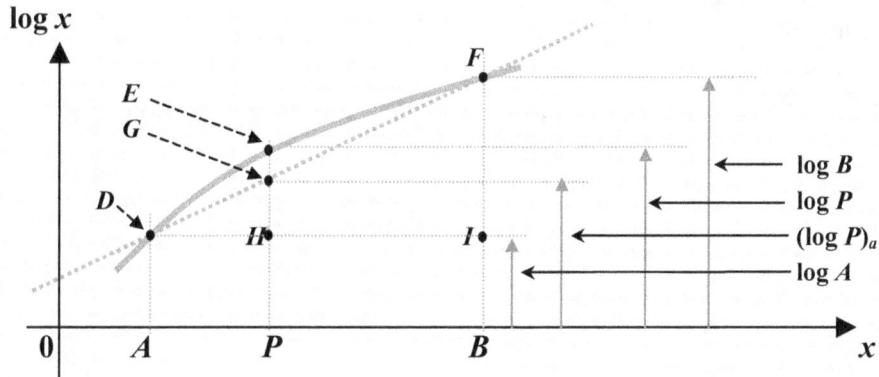

Next, assuming that the distance from **E** to **G** is negligible, which means small enough, we can get an approximate value of **log P** using ratios that apply to similar triangles.

In the graph above, we have two triangles **DIF** and **DHG** similar to each other, because $\overline{FI}$ and $\overline{GH}$ are parallel to each other.    Therefore, we can see that:

$$\overline{AB} : \overline{AP} = \overline{DI} : \overline{DH} = \overline{FI} : \overline{GH} \Rightarrow \overline{AB} : \overline{AP} = \overline{FI} : \overline{GH}.$$

That is to say that assuming **AB** is the distance from **A** to **B**, we can get:

$$\frac{AB}{AP} = \frac{DI}{DH} = \frac{FI}{GH} \Rightarrow \frac{AB}{AP} = \frac{FI}{GH}.$$

So next, assuming for simplicity, $d = GH$, we can see that $\log A + d = (\log P)_a$, which is an approximation of **log P**. And thus, we want to find **d**.

To begin with, we can get: $\dfrac{AB}{AP} = \dfrac{FI}{d} \Rightarrow d \cdot \dfrac{AB}{AP} = FI \Rightarrow d = FI \cdot \dfrac{AP}{AB}$.

And next, we can get: $FI = \log B - \log A$, $AP = P - A$, and $AB = B - A$.

So we get: $d = (\log B - \log A) \cdot \dfrac{P - A}{B - A}$.  And we have:

$A = 5.72$, $\log A = 0.7574$, $B = 5.73$, $\log B = 0.7582$, and $P = 5.724$.

So first, we can get:

$P - A = 5.724 - 5.72 = 0.004$, $\log B - \log A = 0.7582 - 0.7574 = 0.0008$, and

$B - A = 5.73 - 5.72 = 0.01$.

So next, we can get:

$d = \dfrac{P - A}{B - A} \cdot (\log B - \log A) = \dfrac{0.004}{0.01} \cdot 0.0008 = 0.4 \cdot 0.0008 = 0.00032$.

Now, we have: $\log A + d = (\log P)_a$, which is an approximation of $\log P = \log 5.724$.

So we get: $(\log P)_a = \log A + d = 0.7574 + 0.00032 = 0.75772 \cong 0.7577$.

Therefore, $\log 5.724 \cong 0.7577$, which is in fact, just about the same as the value we can get using a calculator, too.

**In short:**

Setting $A = 5.72$, and $B = 5.73$, and putting all the values in a graph, we can get:

**Fig. 0.2**        **log 5.72 = 0.7574, log 5.73 = 0.7582, and $P$ = 5.724**

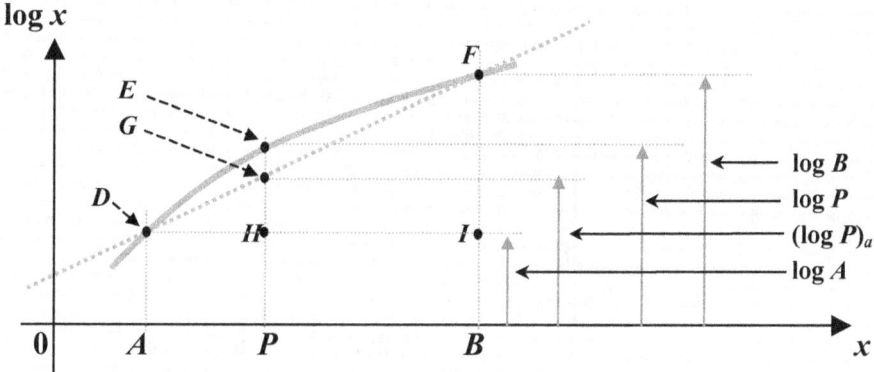

Since **FI** is parallel to **GH**, two triangles **DIF** and **DHG** are similar to each other.

So setting: **GH** = **d**, and letting **(log $P$)$_a$** be the approximation, we can set:

**(log $P$)$_a$ = log $A$ + $d$**, where $d = \frac{P-A}{B-A} \cdot (\log B - \log A)$.

We have: $P - A = 5.724 - 5.72 = 0.004$, $B - A = 5.73 - 5.72 = 0.01$, and

$\log B - \log A = 0.7582 - 0.7574 = 0.0008$.

So we get: $d = \frac{P-A}{B-A} \cdot (\log B - \log A) = \frac{0.004}{0.01} \cdot 0.0008 = 0.4 \cdot 0.0008 = 0.00032$.

Thus, **log 5.724 $\cong$ (log $P$)$_a$ = log $A$ + $d$ = 0.7574 + 0.00032 = 0.75772 $\cong$ 0.7577**.

**Note:**

We don't have to take as a formula, **(log $P$)$_a$ = log $A$ + $d$** where $d = \frac{P-A}{B-A} \cdot (\log B - \log A)$, nor yet to memorize it. Not using it often, we get to forget it soon anyway.

Like any others in math, it's an idea.
So understanding the idea, and doing some examples, we can get used to it. And then, we can readily derive it by simple geometry on similar triangles.
And getting used to the idea, we can calculate separately the two ratios using two right triangles similar to each other:

One is the ratio between differences in antilogs, and the other is the ratio in logs.

For simplicity, we can make a diagram as below.

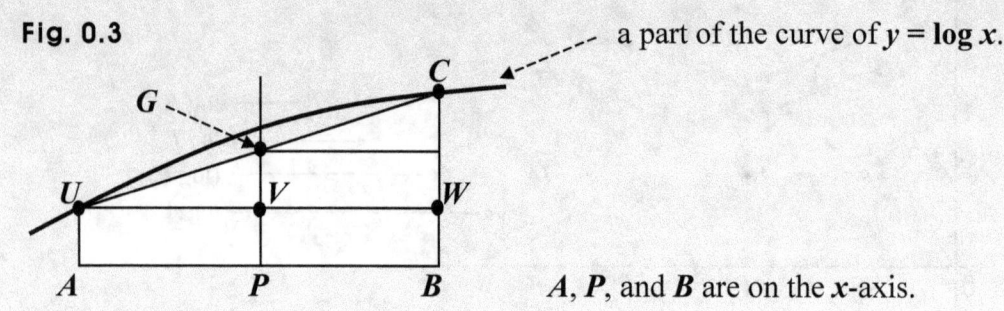

**Fig. 0.3**

a part of the curve of $y = \log x$.

$A$, $P$, and $B$ are on the $x$-axis.

Assuming $UABW$ is a rectangle, and $(\log P)_a$ is the approximation, we can set:

$G = (P, (\log P)_a)$, and $V = (P, \log A)$.

Then, setting: $VG = d$, we get: $\log P \cong (\log P)_a = \log A + d$. So we want to get $d$ now.

In the graph above, we can see two right triangles $UVG$ and $UWC$, which are similar to each other, since $VG$ is parallel to $WC$. So we get:

$$UV : VG = UV : d = UW : WC \Rightarrow \frac{UV}{d} = \frac{UW}{WC} \Rightarrow d = UV \cdot \frac{WC}{UW}.$$

Since $U$ and $C$ are points in the curve, we can set:

$U = (A, \log A)$, and $C = (B, \log B)$.

And since $UABW$ is a rectangle, we get:

$UV = P - A$, $UW = B - A$, and $WC = \log B - \log A$.

And in this example, we have:

$A = 5.72$, $\log A = 0.7574$, $B = 5.73$, $\log B = 0.7582$, and $P = 5.724$.

Then, we get: $UW = B - A = 5.73 - 5.72 = 0.01$, $UV = P - A = 5.724 - 5.72 = 0.004$,

and $WC = \log B - \log A = 0.7582 - 0.7574 = 0.0008$,

So we get: $d = UV \cdot \frac{WC}{UW} = 0.004 \cdot \frac{0.0008}{0.01} = 0.004 \cdot 0.08 = 0.00032$.

Thus, we get: $\log 5.724 \cong (\log P)_a = \log A + d = 0.7574 + 0.00032 = 0.75772 \cong 0.7577$.

## Suggestions or Solutions
## To the Problem in the Example 1

**Using log 5.72 = 0.7574 and log 5.73 = 0.7582, find an approximation of the antilog** *A* **in log** *A* **= 0.7579 with no calculator.**

We have: **log 5.72 = 0.7574, log 5.73 = 0.7582**, and **log $A$ = 0.7579**.

Setting: $M = 5.72$, and $N = 5.73$, we get: **log $M$ = 0.7574**, and **log $N$ = 0.7582**.

Assuming $A_a$ is the approximation, and putting in a graph all the values above, we get:

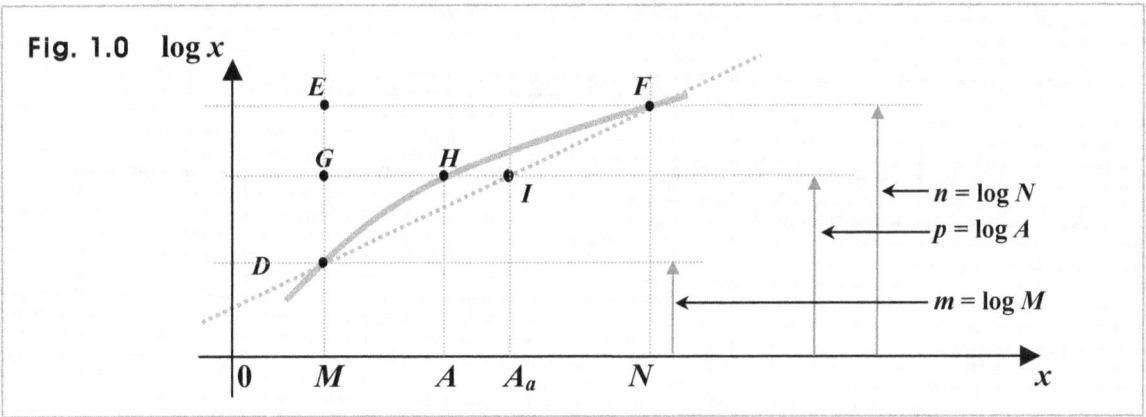

Two triangles **DGI** and **DEF** are similar to each other, because **GI** is parallel to **EF**.

Assuming **GI = d**, we get: $A_a = M + d$, where $d = (N - M) \cdot \dfrac{\log A - \log M}{\log N - \log M}$.

We have: $N - M = 5.73 - 5.72 = 0.01$, **log $A$ – log $M$** = 0.7579 – 0.7574 = 0.0005, and **log $N$ – log $M$** = 0.7582 – 0.7574 = 0.0008.

Thus, we get: $d = (N - M) \cdot \dfrac{\log A - \log M}{\log N - \log M} = 0.01 \cdot \dfrac{0.0005}{0.0008} = \dfrac{0.05}{8} = 0.00625$.

Therefore, $A \cong A_a = M + d \cong 5.72 + 0.00625 = 5.72625 \Rightarrow A \cong 5.7263$.

*If not quite sure of the idea behind the processes above, follow the steps below:*

To begin with, what do we mean by an approximation of an antilog?

By the definition for logs, we get: $A = 10^{0.7579} \Leftrightarrow 0.7579 = \log A$, where $A$ is the antilog.

So this problem is asking us to get an approximate value of the power of 10, which is $10^{0.7579}$.   How then, can we get such an approximation, without a calculator, of course?

Using two ratios that apply to similar triangles, we can get it the way as follows:

Setting first, $M = 5.72$, and $N = 5.73$, we get: $\log M = 0.7574$, and $\log N = 0.7582$.

Next, assuming $A_a$ is an approximation of $A$, and putting in a graph all the values above, we can get one as shown below.

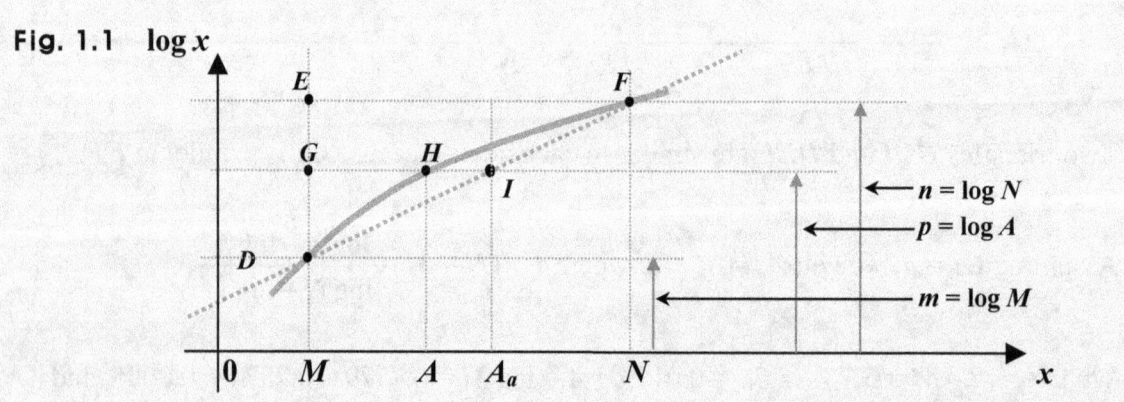

Fig. 1.1   log x

In the graph above, assuming that the distance between $H$ and $I$ is sufficiently small, we can get an approximation of $A$ by means of ratios that apply to similar triangles.
And we can get it the way below:

First, two triangles $DGI$ and $DEF$ are similar to each other, since $GI$ is parallel to $EF$.

So we get: $DG : DE = GI : EF$. That is, we get: $\dfrac{DG}{DE} = \dfrac{GI}{EF}$.

And assuming for simplicity, $d = GI$, we can get:

$M + d = A_a$, which is an approximation of $A$.   So we need to find $d$ now.

First, we can get: $\dfrac{DG}{DE} = \dfrac{GI}{EF} \Rightarrow GI = EF \cdot \dfrac{DG}{DE}$, which is $d$.

And next, we can get: $EF = N - M$, $DG = \log A - \log M$, and $DE = \log N - \log M$.

So we get: $d = (N - M) \cdot \dfrac{\log A - \log M}{\log N - \log M}$.

Now, we put the following values into the equation above:

$M = 5.72$, $\log M = 0.7574$, $N = 5.73$, $\log N = 0.7582$, and $\log A = 0.7579$.

Then, we get: $N - M = 5.73 - 5.72 = 0.01$, $\log A - \log M = 0.7579 - 0.7574 = 0.0005$,

and $\log N - \log M = 0.7582 - 0.7574 = 0.0008$.

So we get: $d = (N - M) \cdot \dfrac{\log A - \log M}{\log N - \log M} = 0.01 \cdot \dfrac{0.0005}{0.0008} = \dfrac{0.05}{8} = 0.00625$.

Thus, we get: $A \cong A_a = M + d \cong 5.72 + 0.00625 = 5.72625$.

Therefore, $A \cong 5.7263$. And in fact, using a calculator, we get: $\log 5.7263 \cong 0.7579$.

**In short:**

We have: **log 5.72 = 0.7574, log 5.73 = 0.7582, and log A = 0.7579.**

Setting: $M = 5.72$, and $N = 5.73$, we get: **log $M$ = 0.7574, and log $N$ = 0.7582.**

Assuming $A_a$ is the approximation, and putting in a graph all the values above, we get:

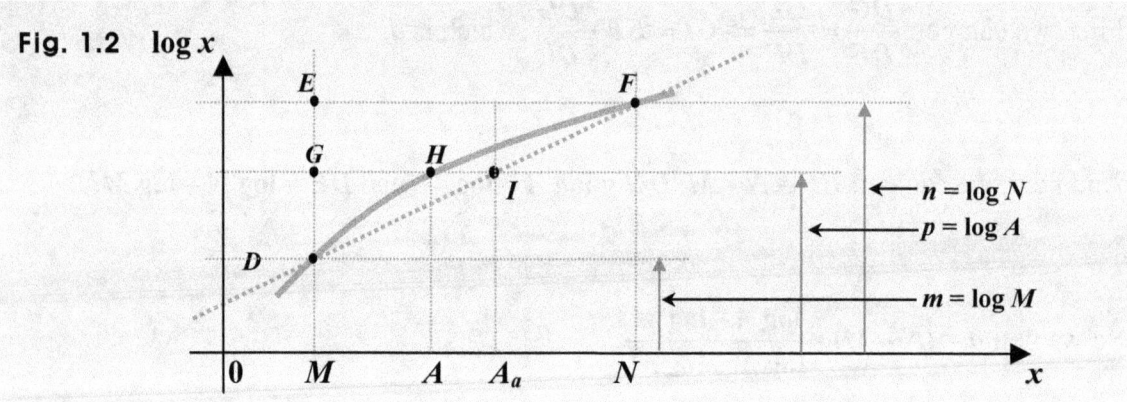

Two triangles **DGI** and **DEF** are similar to each other, because **GI** is parallel to **EF**.

Assuming $GI = d$, we get: $A_a = M + d$, where $d = (N - M) \cdot \dfrac{\log A - \log M}{\log N - \log M}$.

We have: $N - M = 5.73 - 5.72 = 0.01$, **log $A$ − log $M$** = 0.7579 − 0.7574 = 0.0005, and

**log $N$ − log $M$** = 0.7582 − 0.7574 = 0.0008.

Thus, we get: $d = (N - M) \cdot \dfrac{\log A - \log M}{\log N - \log M} = 0.01 \cdot \dfrac{0.0005}{0.0008} = \dfrac{0.05}{8} = 0.00625.$

Therefore, $A \cong A_a = M + d \cong 5.72 + 0.00625 = 5.72625 \Rightarrow A \cong 5.7263.$

**Note:**

We don't have to take for a formula, $A_a = M + d$, where $d = (N - M) \cdot \dfrac{\log A - \log M}{\log N - \log M}$,

nor yet to memorize it. Not using it often, we forget it quickly anyway. Understanding the idea, we can readily derive it by simple geometry on similar triangles.

Getting used to the idea, we can calculate separately two ratios that apply to similar right triangles:

One is the ratio between differences in antilogs, and the other is the ratio in logs.

Solving problems in this kind, we may want to use a diagram as below, where the curve in gray is a part of the curve of $y = \log x$.

**Fig. 1.3**

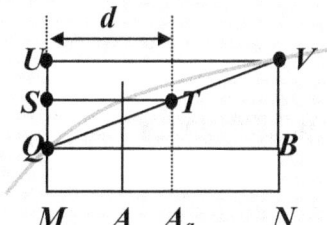

$M, A, A_a$, and $N$ are on the $x$-axis.

Then first, assuming **QMNB** is a rectangle, $A_a$ is an approximation of $A$, $T = (A_a, \log A)$, and $d = ST$, we can get: $A \cong A_a = M + d$.

Next, assuming **BQUV** is a rectangle, we get:

$QU = \log N - \log M$, $UV = N - M$, and $QS = \log A - \log M$.

And next, we can see that two right triangles **QST** and **QUV** are similar to each other.

So we get: $\dfrac{UV}{ST} = \dfrac{QU}{QS}$. And we can say that:

$\dfrac{UV}{ST}$ is the ratio between differences in antilogs.

$\dfrac{QU}{QS}$ is the ratio between differences in logs.

And we know: $d = ST$.   So we get:

$$\frac{UV}{ST} = \frac{QU}{QS} \Rightarrow \frac{UV}{d} = \frac{QU}{QS} \Rightarrow UV = d \cdot \frac{QU}{QS} \Rightarrow d = UV \cdot \frac{QS}{QU}.$$

Now, in this example, we have:

$M = 5.72$, $\log M = 0.7574$, $N = 5.73$, $\log N = 0.7582$, and $\log A = 0.7579$.

So putting the information above in the diagram, we get.

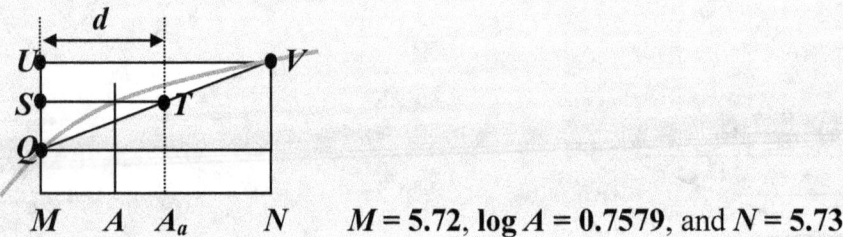

**Fig. 1.4**

$M = 5.72$, $\log A = 0.7579$, and $N = 5.73$.

Then, taking the ratio between differences in antilogs, we get first:

$UV = N - M = 5.73 - 5.72 = 0.01$, and $ST = d$, so the ratio is: $\dfrac{UV}{d} = \dfrac{0.01}{d}$.

Next, taking the ratio between differences in logs, we get first:

$QU = \log N - \log M = 0.0008$, and $QS = \log A - \log M = 0.0005$.

So the ratio is: $\dfrac{QU}{QS} = \dfrac{0.0008}{0.0005} = \dfrac{8}{5}$.

And we have: $\dfrac{UV}{d} = \dfrac{QU}{QS}$.

So we get: $\dfrac{0.01}{d} = \dfrac{8}{5} \Rightarrow d = 0.01 \cdot \dfrac{5}{8} = \dfrac{5}{800} = 0.00625$.

Thus, we get: $A \cong A_a = M + d \cong 5.72 + 0.00625 = 5.72625 \Rightarrow A \cong 5.7263$.

## Examples 2 in Approximations

Getting an approximation of a log-value, we need to have two neighboring log-values sufficiently close to each other. What if we don't have such two?

There must be such two log-values if the problem is well defined, that is, solvable. So if not shown, they are probably hidden somewhere in the problem, in other words, implied. Somehow thus, we should be able to manage to find them. And then, we use ratios that apply to similar triangles. And the same is true for an approximation of an antilog, too.

Doing the examples below, use no calculator.

0.   Assuming **log 8 = $p$**, and **log 81 = $q$**:

0.0   Find an approximation of **log 8.4**.

0.1.   Find an approximation of **$0.84^5$**.

1.   Using the values below:

**log 5.67 = 0.7536, log 5.68 = 0.7543, log 4.82 = 0.6830, log 4.83 = 0.6839,**
**log 13.8 = 1.1399, and log 13.9 = 1.1430,**

find an approximate value of $\sqrt[3]{\frac{0.5673^2(48.26)}{0.1382}}$.

## Suggestions or Solutions
## To the Problem 0 in the Example 0

**Assuming log 8 = $p$, and log 81 = $q$, find an approximation of log 8.4**

We can have: **log 8 < log 8.4 < log 9.**    And we have: **log 81 = $q$.**

So we get: **log 81 = log $9^2$ = 2 log 9 = $q$ $\Rightarrow$ log 9 = 0.5$q$.**

And below is a diagram that has a part of the curve of $y$ = **log** $x$.

**Fig. 0.0**

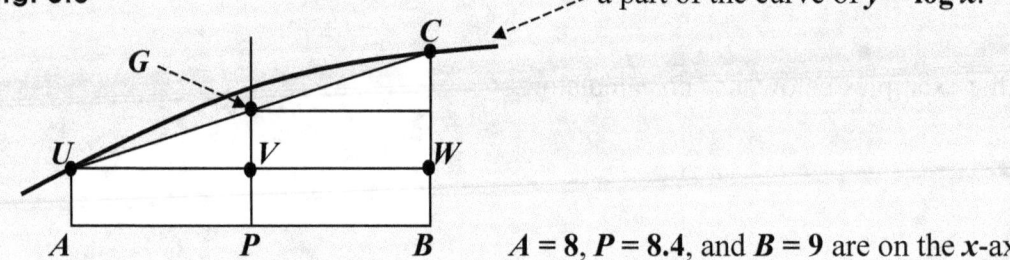

a part of the curve of $y$ = **log** $x$.

$A$ = 8, $P$ = 8.4, and $B$ = 9 are on the $x$-axis.

Assuming $UABW$ is a rectangle, and **(log $P$)**$_a$ is the approximation, we can set:

$G$ = ($P$, **(log $P$)**$_a$), and $V$ = ($P$, **log** $A$).

Then, setting: $VG$ = $d$, we get: **log** $P \cong$ **(log $P$)**$_a$ = **log** $A$ + $d$.

In the graph above, we can see two right triangles $UVG$ and $UWC$, which are similar to each other, since $VG$ is parallel to $WC$. So we get:

$UV : VG = UV : d = UW : WC \Rightarrow \frac{UV}{d} = \frac{UW}{WC} \Rightarrow d = UV \cdot \frac{WC}{UW}$.

Since $UABW$ is a rectangle, and $U$ and $C$ are two points in the curve, we can set:

$WC$ = **log** $B$ – **log** $A$ = **log** 9 – **log** 8 = 0.5$q$ – $p$,

$UW$ = $B$ – $A$ = 9 – 8 = 1, and $UV$ = $P$ – $A$ = 8.4 – 8 = 0.4.

So we get: $d = UV \cdot \frac{WC}{UW} = 0.4 \cdot \frac{0.5q-p}{1} = 0.2q - 0.4p$.

And thus, **log 8.4** $\cong$ **(log P)$_a$** = **log A + d = log 8 + d = p + 0.2q − 0.4p = 0.6p + 0.2q**.

*If not quite sure of the idea behind the processes above, follow the steps below:*

We need two neighboring log-values to **log 8.4**, and they can be **log 8** and **log 9**.

We have the value of **log 8**, but the value of **log 9** is not given, which is not quite the case though. Why not?

We can get the value of **log 9**, because we are given the value of **log 81**.
How then, can we get the value of **log 9**?

We know: **log 81 = log 9$^2$ = 2 log 9**, and we have: **log 81 = q**.
So we get: **2 log 9 = q** $\Rightarrow$ **log 9 = 0.5q**.

Now that we've got the value of **log 9**, we can get an approximate value of **log 8.4**.

Setting first: $A = 8$, $P = 8.4$, and $B = 9$, we can make a diagram as below:

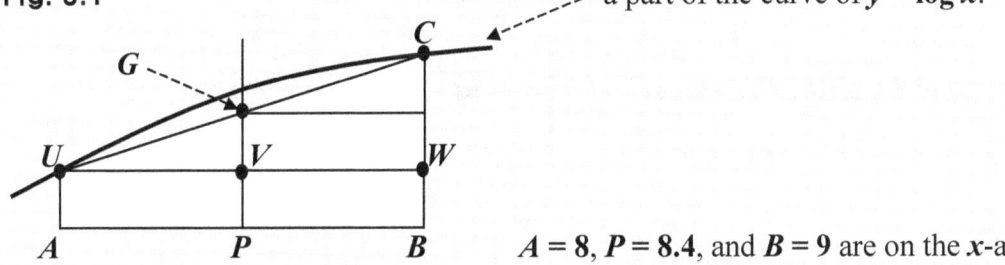

**Fig. 0.1**                                              a part of the curve of $y = \log x$.

$A = 8$, $P = 8.4$, and $B = 9$ are on the $x$-axis.

So next, assuming $UABW$ is a rectangle, and **(log P)$_a$** is the approximation, we can set:

$G = (P, (\log P)_a)$, and $V = (P, \log A)$.

Then, setting: $VG = d$, we get: $\log P \cong (\log P)_a = \log A + d$.   So we want to get $d$ now.

In the diagram above, we can see two right triangles $UVG$ and $UWC$, which are similar to each other, since $VG$ is parallel to $WC$. So we get:

$$UV : VG = UV : d = UW : WC \Rightarrow \frac{UV}{d} = \frac{UW}{WC} \Rightarrow UV = \frac{UW}{WC} \cdot d \Rightarrow d = UV \cdot \frac{WC}{UW}.$$

Then, first, since $U$ and $C$ are points in the curve, we can set:

$U = (A, \log A)$, and $C = (B, \log B)$.

Next, since $UABW$ is a rectangle, we get:

$UV = P - A$, $UW = B - A$, and $WC = \log B - \log A$.

And in this example, we have:

$A = 8$, $\log A = p$, $B = 9$, $\log B = 0.5q$, and $P = 8.4$.

Then, we get: $UW = B - A = 9 - 8 = 1$, $UV = P - A = 8.4 - 8 = 0.4$,

and $WC = \log B - \log A = 0.5q - p$.

So we get: $d = UV \cdot \frac{WC}{UW} = 0.4 \cdot \frac{0.5q - p}{1} \Rightarrow d = 0.4(0.5q - p) = 0.2q - 0.4p$.

And thus, $\log 8.4 \cong (\log P)_a = \log A + d = p + 0.2q - 0.4p = 0.6p + 0.2q$.

**In short:**

We can have: $\log 8 < \log 8.4 < \log 9$. And we have: $\log 81 = q$.

So we get: $\log 81 = \log 9^2 = 2 \log 9 = q \Rightarrow \log 9 = 0.5q$.

And below is a diagram that has a part of the curve of $y = \log x$.

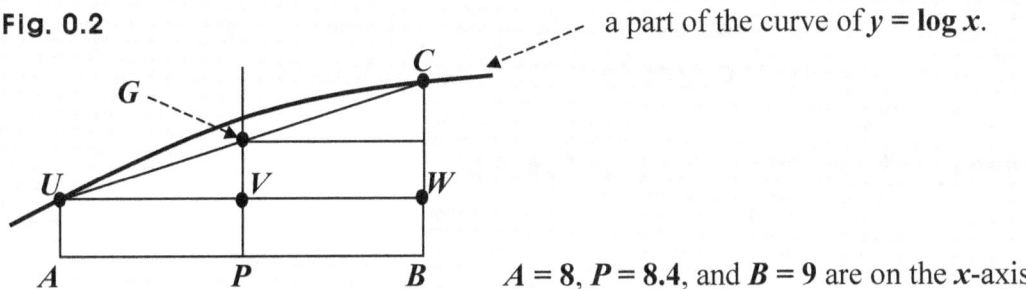

**Fig. 0.2**    a part of the curve of $y = \log x$.

$A = 8$, $P = 8.4$, and $B = 9$ are on the $x$-axis.

Assuming **UABW** is a rectangle, and **$(\log P)_a$** is the approximation, we can set:

$G = (P, (\log P)_a)$, and $V = (P, \log A)$.

Then, setting: $VG = d$, we get: $\log P \cong (\log P)_a = \log A + d$.

In the graph above, we can see two right triangles **UVG** and **UWC**, which are similar to each other, since **VG** is parallel to **WC**. So we get:

$UV : VG = UV : d = UW : WC \Rightarrow \frac{UV}{d} = \frac{UW}{WC} \Rightarrow d = UV \cdot \frac{WC}{UW}$.

Since **UABW** is a rectangle, and **U** and **C** are two points in the curve, we can set:

$WC = \log B - \log A = \log 9 - \log 8 = 0.5q - p$,

$UW = B - A = 9 - 8 = 1$, and $UV = P - A = 8.4 - 8 = 0.4$.

So we get: $d = UV \cdot \frac{WC}{UW} = 0.4 \cdot \frac{0.5q-p}{1} = 0.2q - 0.4p$.

And thus, $\log 8.4 \cong (\log P)_a = \log A + d = \log 8 + d = p + 0.2q - 0.4p = 0.6p + 0.2q$.

## Suggestions or Solutions
## To the Problem 1 in the Example 0

**Assuming log 8 = *p*, and log 81 = *q*, find an approximation of 0.84$^5$.**

**log 0.84$^5$ = 5 log 0.84 = 5 log $\frac{8.4}{10}$ = 5(log 8.4 – log 10) = 5(log 8.4 – 1)**

**≅ 5(0.6*p* + 0.2*q* – 1) = 3*p* + *q* – 5.** And we have: **log 8 = *p*, and log 81 = *q*.**

So we get: **log 0.84$^5$ ≅ 3*p* + *q* – 5 = 3 log 8 + log 81 – 5**

**= log 512 + log 81 – 5 log 10 = log (512·81) – 5 log 10 = log 41472 + log 10$^{-5}$**

**= log (41472·10$^{-5}$) = log 0.41472.** Therefore, **0.84$^5$ ≅ 0.41472.**

*If not quite sure of the idea behind the processes above, follow the steps below:*

For this example, we can use the approximation of **log 8.4** found above (Problem 0).

Let's first, express **log 0.84$^5$** using **log 8.4**.    Then, we get:

**log 0.84$^5$ = 5 log 0.84 = 5 log $\frac{8.4}{10}$ = 5(log 8.4 – log 10) = 5(log 8.4 – 1).**

And we know: **log 8.4 ≅ 0.6*p* + 0.2*q*.**

So we get: **log 0.84$^5$ ≅ 5(0.6*p* + 0.2*q* – 1) = 3*p* + *q* – 5.**

And we have: **log 8 = *p*, and log 81 = *q*.**

So we get: **log 0.84$^5$ ≅ 3*p* + *q* – 5 = 3 log 8 + log 81 – 5.**

And we have: $\mathbf{3 \log 8 = \log 8^3 = \log 512}$, and $\mathbf{5 = 5 \log 10}$.

So we get: $\mathbf{3 \log 8 + \log 81 - 5 = \log 512 + \log 81 - 5 \log 10 = \log (512{\cdot}81) - 5 \log 10}$

$\mathbf{= \log 41472 + (\text{-}5) \log 10 = \log 41472 + \log 10^{-5} = \log (41472{\cdot}10^{-5}) = \log 0.41472}$.

Therefore, $\mathbf{0.84^5 \cong 0.41472}$. Using a calculator, we get: $\mathbf{0.84^5 = 0.4182119424}$.

**In short:**

$\mathbf{\log 0.84^5 = 5 \log 0.84 = 5 \log \frac{8.4}{10} = 5(\log 8.4 - \log 10) = 5(\log 8.4 - 1)}$

$\mathbf{\cong 5(0.6p + 0.2q - 1) = 3p + q - 5}$. And we have: $\mathbf{\log 8 = p}$, and $\mathbf{\log 81 = q}$.

So we get: $\mathbf{\log 0.84^5 \cong 3p + q - 5 = 3 \log 8 + \log 81 - 5}$

$\mathbf{= \log 512 + \log 81 - 5 \log 10 = \log (512{\cdot}81) - 5 \log 10 = \log 41472 + \log 10^{-5}}$

$\mathbf{= \log (41472{\cdot}10^{-5}) = \log 0.41472}$. Therefore, $\mathbf{0.84^5 \cong 0.41472}$.

## Suggestions or Solutions
## To the Problem in the Example 1

**We have:**

**log 5.72 = 0.7574, log 5.73 = 0.7582, log 3.58 = 0.5539, log 3.59 = 0.5551,**

**log 29.5 = 1.4698, log 29.6 = 1.4713, log 3.30 = 0.5185, and log 3.31 = 0.5198.**

**Now, find an approximate value of $\sqrt[5]{\frac{0.5724^2 \cdot 3583}{2.953}}$ .**

All the values given are log-values.
And solving this problem, we are going to make use of them, of course.

However, the value we want to get the approximation of is not a log-value. So?

We may want to put it in a common log. That is, we take the log of $\sqrt[5]{\frac{0.5724^2 \cdot 3583}{2.953}}$ .

Then, we get: $\log \sqrt[5]{\frac{0.5724^2 \cdot 3583}{2.953}}$ , the value of which is a log-value. So what?

We can use some of the values given so that we can get an approximate value of the log above. And after taking the approximation, we can get the approximation of the antilog, which is: $\sqrt[5]{\frac{0.5724^2 \cdot 3583}{2.953}}$ . How?

We find first, the approximate value of $\log \sqrt[5]{\frac{0.5724^2 \cdot 3583}{2.953}}$ using the log values given, and then, using the approximate value found, we get the approximate value of $\sqrt[5]{\frac{0.5724^2 \cdot 3583}{2.953}}$ .

So let's now, get the approximation of the log value first.

Then, setting first: $r = \sqrt[5]{\frac{0.5724^2 \cdot 3583}{2.953}}$, and taking the common logs of both sides, we get:

$$\log r = \log\{\tfrac{0.5724^2 \cdot 3583}{2.953}\}^{\frac{1}{5}} = \tfrac{1}{5}\log\{\tfrac{0.5724^2 \cdot 3583}{2.953}\} = \tfrac{1}{5}\{\log(0.5724^2 \cdot 3583) - \log 2.953\}$$

$$= \tfrac{1}{5}\{(\log 0.5724^2 + \log 3583) - \log 2.953\} = \tfrac{1}{5}(2\log 0.5724 + \log 3583 - \log 2.953).$$

Now, in the expression above, we have: **log 0.5727**, **log 3583**, and **log 2.953**, the values of which are however, not given.    Then, what are we going to do with those?

We want to take approximations of those. However, we are not given values close to them, that is, we are not given values neighboring each of those.

We are given values though, close to each of **log 5.724**, **log 3.583**, and **log 29.53**.

For instance, for **log 5.724**, we are given: **log 5.72 = 0.7574**, and **log 5.73 = 0.7582**.

So let's now, begin with getting an approximation of **log 5.724**.

Then first, we can set up a diagram as follows:

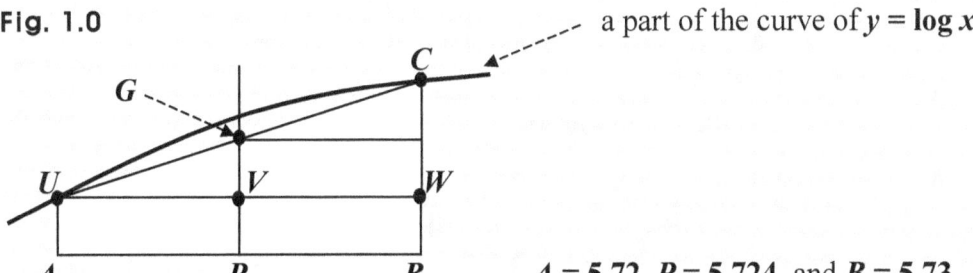

**Fig. 1.0**                                   a part of the curve of $y = \log x$.

$A = 5.72$, $P = 5.724$, and $B = 5.73$.

Assuming **UABW** is a rectangle, **(log P)$_a$** is the approximation, and **VG = d**, we can set:

$\log P \cong (\log P)_a = \log A + d$.    So we want to get **d** now.

In the graph above, we can see two right triangles **UVG** and **UWC**, which are similar to each other, since **VG** is parallel to **WC**.    So we get:

$$UV : VG = UV : d = UW : WC \Rightarrow \tfrac{UV}{d} = \tfrac{UW}{WC} \Rightarrow d = UV \cdot \tfrac{WC}{UW}.$$

And since *UABW* is a rectangle, we get:

$UV = P - A$, $UW = B - A$, and $WC = \log B - \log A$.

And in this case, we have:

$A = 5.72$, $\log A = 0.7574$, $B = 5.73$, $\log B = 0.7582$, and $P = 5.724$.

Then, we get: $UW = B - A = 5.73 - 5.72 = 0.01$, $UV = P - A = 5.724 - 5.72 = 0.004$,

and $WC = \log B - \log A = 0.7582 - 0.7574 = 0.0008$.

So we get: $d = UV \cdot \frac{WC}{UW} = 0.004 \cdot \frac{0.0008}{0.01} = 0.004 \cdot 0.08 = 0.00032$.

Thus, we get: $\log 5.724 \cong (\log P)_a = \log A + d = 0.7574 + 0.00032 = 0.75772 \cong 0.7577$.

Therefore, we get: $\log 0.5724 \cong -1 + 0.7577 = \overline{1}.7577$.   Why -1, though?

The highest place value in 0.5724 is $10^{-1}$, and -1 is the characteristic.

Now, by the same token, we can get an approximation of **log 3.583**, and then, **log 3583**.

We have: **log 3.58 = 0.5539**, and **log 3.59 = 0.5551**.

Then again, we can set up a diagram as follows:

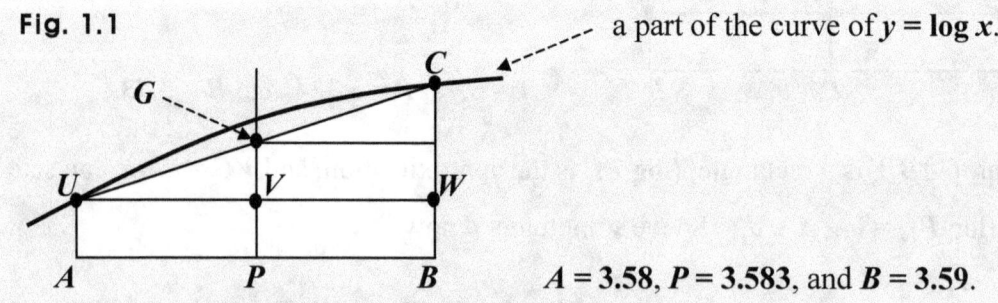

**Fig. 1.1**   a part of the curve of $y = \log x$.

$A = 3.58$, $P = 3.583$, and $B = 3.59$.

Assuming first, **(log P)_a** is the approximation, and $VG = d$, we can set:

$\log P \cong (\log P)_a = \log A + d$.   So we want to get *d* now.

Since two right triangles *UVG* and *UWC* are similar to each other, we get:

$$UV : VG = UV : d = UW : WC \Rightarrow \frac{UV}{d} = \frac{UW}{WC} \Rightarrow d = UV \cdot \frac{WC}{UW}.$$

And since *UABW* is a rectangle, we get:

$UV = P - A$, $UW = B - A$, and $WC = \log B - \log A$.

And in this case, we have:

$A = 3.58$, $\log A = 0.5539$, $B = 3.59$, $\log B = 0.5551$, and $P = 3.583$.

Then, we get: $UW = B - A = 3.59 - 3.58 = 0.01$, $UV = P - A = 3.583 - 3.58 = 0.003$, and $WC = \log B - \log A = 0.5551 - 0.5539 = 0.0012$.

So we get: $d = UV \cdot \frac{WC}{UW} = 0.003 \cdot \frac{0.0012}{0.01} = 0.003 \cdot 0.12 = 0.00036$.

Thus, we get: $\log 3.583 \cong (\log P)_a = \log A + d = 0.5539 + 0.00036 = 0.55426 \cong 0.5543$.

Therefore, we get: $\log 3538 \cong 3 + 0.5543 = 3.5543$.   Why 3 though, above the point?

The highest place value in 3538 is $10^3$, and 3 is the characteristic.

Likewise, we can get an approximation of **log 29.53**, and then, **log 2.953**.

We have **log 29.5 = 1.4698**, and **log 29.6 = 1.4713**.

Then again, we can set up a diagram as follows:

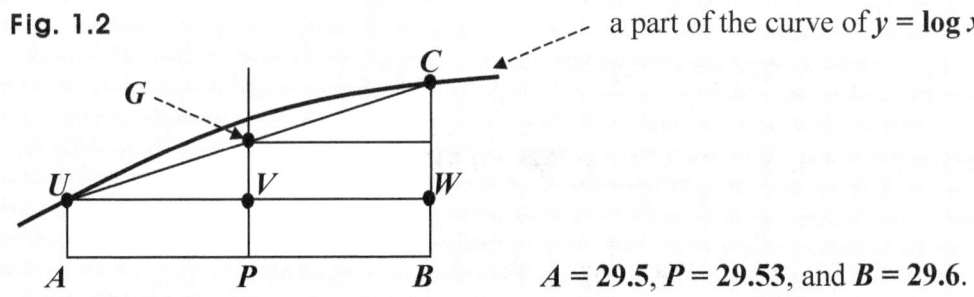

**Fig. 1.2**          a part of the curve of $y = \log x$.

$A = 29.5$, $P = 29.53$, and $B = 29.6$.

174

Assuming first, $(\log P)_a$ is the approximation, and $VG = d$, we can set:

$\log P \cong (\log P)_a = \log A + d$. So we want to get $d$ now.

Next, $UVG$ and $UWC$ are similar triangles, so we get: $\frac{UV}{d} = \frac{UW}{WC} \Rightarrow d = UV \cdot \frac{WC}{UW}$.

And since $UABW$ is a rectangle, we get:

$UV = P - A$, $UW = B - A$, and $WC = \log B - \log A$.

And in this case, we have:

$A = 29.5$, $\log A = 1.4698$, $B = 29.6$, $\log B = 1.4713$, and $P = 29.53$.

Then, we get: $UW = B - A = 29.6 - 29.5 = 0.01$, $UV = P - A = 29.53 - 29.5 = 0.03$,

and $WC = \log B - \log A = 1.4713 - 1.4698 = 0.0015$.

So we get: $d = UV \cdot \frac{WC}{UW} = 0.03 \cdot \frac{0.0015}{0.01} = 0.03 \cdot 0.15 = 0.0045$.

Thus, we get: $\log 29.53 \cong (\log P)_a = \log A + d = 1.4698 + 0.0045 = 1.4743$.

Therefore, we get: $\log 2.953 \cong 0.4743$.   Why 0 though, above the point?

The highest place value in 2.953 is $10^0$, and 0 is the characteristic.

Now, we have: $\log r = \frac{1}{5}(2 \log 0.5724 + \log 3583 - \log 2.953)$.

And we have: $\log 0.5724 \cong \bar{1}.7577$, $\log 3538 \cong 3.5543$, and $\log 2.953 \cong 0.4743$.

So we get: $\log r \cong \frac{1}{5}\{2(-1 + 0.7577) + (3 + 0.5543) - 0.4743\}$

$= \frac{1}{5}(-2 + 1.5154 + 3 + 0.5543 - 0.4743) = \frac{2.5954}{5} = 0.51908 \cong 0.5191$.

Thus, we get: $\log r \cong 0.5191$.   What then, is the next?

Finally, we want to take an approximation of *r*, which is: $\sqrt[5]{\frac{0.5724^2 \cdot 3583}{2.953}}$.

We have: **log 3.30 = 0.5185, log 3.31 = 0.5198**, and **log $r \cong$ 0.5191**.

So we are now taking an approximation of an antilog.

To begin with, setting *r* = *A*, we can set up a diagram as below, where the curve in gray is a part of the curve of *y* = **log** *x*.

**Fig. 1.3**

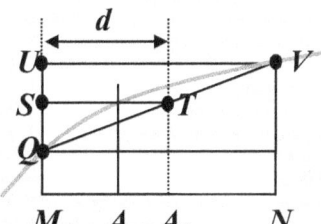

*M*, *A*, $A_a$, and *N* are on the *x*-axis.

Then first, assuming **QMNB** is a rectangle, $A_a$ is an approximation of *A*, *T* = ($A_a$, log *A*), and *d* = *ST*, we can get: $A \cong A_a = M + d$.

Next, assuming **BQUV** is a rectangle, we get:

*QU* = **log** *N* − **log** *M*, *UV* = *N* − *M*, and *QS* = **log** *A* − **log** *M*.

And next, we can see that two right triangles **QST** and **QUV** are similar to each other.

So we get: $\dfrac{UV}{ST} = \dfrac{QU}{QS}$.　　And we can say that:

$\dfrac{UV}{ST}$ is the ratio between differences in antilogs.

$\dfrac{QU}{QS}$ is the ratio between differences in logs.

And we know: $d = ST$. So we get:

$$\frac{UV}{ST} = \frac{QU}{QS} \Rightarrow \frac{UV}{d} = \frac{QU}{QS} \Rightarrow UV = d \cdot \frac{QU}{QS} \Rightarrow d = UV \cdot \frac{QS}{QU}.$$

Now, in this example, we have:

$M = 3.30$, $\log M = 0.5185$, $N = 3.31$, $\log N = 0.5198$, and $\log A = 0.5191$.

So putting the information above in the diagram, we get:

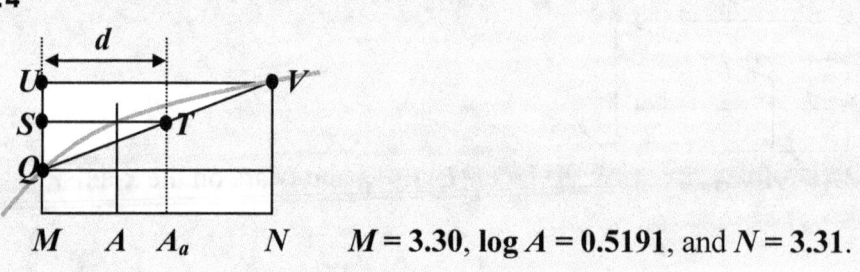

**Fig. 1.4**

$M = 3.30$, $\log A = 0.5191$, and $N = 3.31$.

Then, taking the ratio between differences in antilogs, we get first:

$UV = N - M = 3.31 - 3.30 = 0.01$, and $ST = d$, so the ratio is: $\dfrac{UV}{d} = \dfrac{0.01}{d}$.

Next, taking the ratio between differences in logs, we get first:

$QU = \log N - \log M = 0.0013$, and $QS = \log A - \log M = 0.0006$.

So the ratio is: $\dfrac{QU}{QS} = \dfrac{0.0013}{0.0006} = \dfrac{13}{6}$. And we have: $\dfrac{UV}{d} = \dfrac{QU}{QS}$.

So we get: $\dfrac{0.01}{d} = \dfrac{13}{6} \Rightarrow d = 0.01 \cdot \dfrac{6}{13} = \dfrac{6}{1300} = 0.00462$.

Thus, we get: $A \cong A_a = M + d \cong 3.30 + 0.00462 = 3.30462$

$\Rightarrow A = r = \sqrt[5]{\dfrac{0.5724^2 \cdot .3583}{2.953}} \cong 3.3046$. Using a calculator, we get: $\sqrt[5]{\dfrac{0.5724^2 \cdot .3583}{2.953}} \cong 3.3104$.

**In short:**

Setting first: $r = \sqrt[5]{\frac{0.5724^2 \cdot 3583}{2.953}}$, and taking the common logs of both sides, we get:

$$\log r = \log\{\tfrac{0.5724^2 \cdot 3583}{2.953}\}^{\frac{1}{5}} = \tfrac{1}{5}\log\{\tfrac{0.5724^2 \cdot 3583}{2.953}\} = \tfrac{1}{5}\{\log(0.5724^2 \cdot 3583) - \log 2.953\}$$

$$= \tfrac{1}{5}\{(\log 0.5724^2 + \log 3583) - \log 2.953\} = \tfrac{1}{5}(2\log 0.5724 + \log 3583 - \log 2.953).$$

Let's next, get an approximation of **log 5.724**, and then, **log 0.5724**.
Then first, we can set up a diagram as follows:

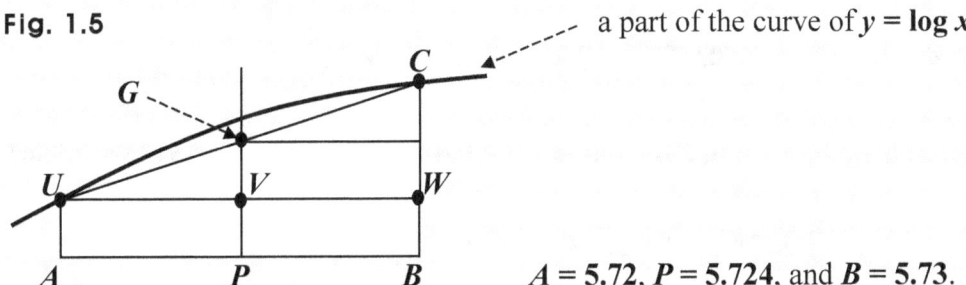

**Fig. 1.5**     a part of the curve of $y = \log x$.

$A = 5.72$, $P = 5.724$, and $B = 5.73$.

Assuming **UABW** is a rectangle, $(\log P)_a$ is the approximation, and $VG = d$, we can set:
$\log P \cong (\log P)_a = \log A + d.$

Since **UVG** and **UWC** are similar triangles, we get: $\frac{UV}{d} = \frac{UW}{WC} \Rightarrow d = UV \cdot \frac{WC}{UW}.$

And since **UABW** is a rectangle, we get:

$UV = P - A = 5.724 - 5.72 = 0.004,$
$UW = B - A = 5.73 - 5.72 = 0.01$, and $WC = \log B - \log A = 0.7582 - 0.7574 = 0.0008.$

So we get: $d = UV \cdot \frac{WC}{UW} = 0.004 \cdot \frac{0.0008}{0.01} = 0.004 \cdot 0.08 = 0.00032.$

Thus, we get: $\log 5.724 \cong (\log P)_a = \log A + d = 0.7574 + 0.00032 = 0.75772 \cong 0.7577.$

Therefore, we get: $\log 0.5724 \cong -1 + 0.7577 = \bar{1}.7577.$

Next, by the same token, we can get an approximation of **log 3.583**, and then, **log 3583**.

We have: **log 3.58 = 0.5539**, and **log 3.59 = 0.5551**.

Then again, uisng the diagram above, taking **(log $P$)$_a$** as the approximation, and assuming $VG = d$, we can set: $\log P \cong (\log P)_a = \log A + d$.

And we get: $\frac{UV}{d} = \frac{UW}{WC} \Rightarrow d = UV \cdot \frac{WC}{UW}$.

And in this case, we have:

$A = 3.58$, $\log A = 0.5539$, $B = 3.59$, $\log B = 0.5551$, and $P = 3.583$.

Then, $UW = B - A = 0.01$, $UV = P - A = 0.003$, and $WC = \log B - \log A = 0.0012$.

So we get: $d = UV \cdot \frac{WC}{UW} = 0.003 \cdot \frac{0.0012}{0.01} = 0.003 \cdot 0.12 = 0.00036$.

Thus, we get: $\log 3.583 \cong (\log P)_a = \log A + d = 0.5539 + 0.00036 = 0.55426 \cong 0.5543$.

Therefore, we get: $\log 3538 \cong 3 + 0.5543 = 3.5543$.

Likewise, we can get an approximation of **log 29.53**, and then, **log 2.953**.

And in this case, we have:

$A = 29.5$, $\log A = 1.4698$, $B = 29.6$, $\log B = 1.4713$, and $P = 29.53$.

Then, $UW = B - A = 0.01$, $UV = P - A = 0.03$, and $WC = \log B - \log A = 0.0015$.

So we get: $d = UV \cdot \frac{WC}{UW} = 0.03 \cdot \frac{0.0015}{0.01} = 0.03 \cdot 0.15 = 0.0045$.

Thus, we get: $\log 29.53 \cong (\log P)_a = \log A + d = 1.4698 + 0.0045 = 1.4743$.

Therefore, we get: $\log 2.953 \cong 0.4743$.

Now, we have: $\log r = \frac{1}{5}(2 \log 0.5724 + \log 3583 - \log 2.953)$.

And we have: $\log 0.5724 \cong \overline{1}.7577$, $\log 3538 \cong 3.5543$, and $\log 2.953 \cong 0.4743$.

So we get: $\log r \cong \frac{1}{5}\{2(-1 + 0.7577) + (3 + 0.5543) - 0.4743\} = \frac{2.5954}{5} = 0.51908 \cong 0.5191$.

Thus, we get: $\log r \cong 0.5191$.

So we can now take an approximation of **r**, which is: $\sqrt[5]{\frac{0.5724^2 \cdot .3583}{2.953}}$.

We have: $\log 3.30 = 0.5185$, $\log 3.31 = 0.5198$, and $\log r \cong 0.5191$.

So we are now taking an approximation of an antilog.

To begin with, setting $r = A$, we can set up a diagram as below, where the curve in gray is a part of the curve of $y = \log x$.

**Fig. 1.6**

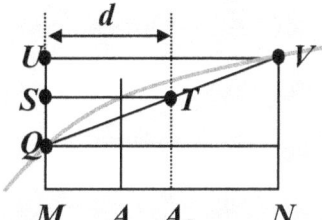

$M, A, A_a$, and $N$ are on the $x$-axis.

Then first, assuming **QMNB** is a rectangle, $A_a$ is an approximation of $A$, $T = (A_a, \log A)$, and $d = ST$, we can get: $A \cong A_a = M + d$.

And next, since **QST** and **QUV** are similar triangles, we get: $d = UV \cdot \dfrac{QS}{QU}$.

Now, in this example, we have:

$M = 3.30$, $\log M = 0.5185$, $N = 3.31$, $\log N = 0.5198$, and $\log A = 0.5191$.

Then, taking the ratio between differences in antilogs, we get first:

$UV = N - M = 3.31 - 3.30 = 0.01$, and $ST = d$, so the ratio is: $\dfrac{UV}{d} = \dfrac{0.01}{d}$.

Next, taking the ratio between differences in logs, we get first:

$QU = \log N - \log M = 0.0013$, and $QS = \log A - \log M = 0.0006$.

So the ratio is: $\dfrac{QU}{QS} = \dfrac{0.0013}{0.0006} = \dfrac{13}{6}$.  And we have: $\dfrac{UV}{d} = \dfrac{QU}{QS}$.

So we get: $\dfrac{0.01}{d} = \dfrac{13}{6} \Rightarrow d = 0.01 \cdot \dfrac{6}{13} = \dfrac{6}{1300} = 0.00462$.

Thus, we get: $A \cong A_a = M + d \cong 3.30 + 0.00462 = 3.30462$

$\Rightarrow A = r = \sqrt[5]{\dfrac{0.5724^2 \cdot 3583}{2.953}} \cong 3.3046$.

Using a calculator, we get: $\sqrt[5]{\dfrac{0.5724^2 \cdot 3583}{2.953}} \cong 3.3104$.

www.ingramcontent.com/pod-product-compliance
Lightning Source LLC
Chambersburg PA
CBHW081119170526
45165CB00008B/2488